Groundwater Geophysics in Hard Rock

Groundwater Geophysics in Hard Rock

Editor

Gyan Singh

scitus academics

Groundwater Geophysics in Hard Rock
Edited by **Gyan Singh**

ISBN: 978-1-68117-680-2
Library of Congress Control Number: 2016958833

SCITUS Academics LLC,
616, Corporate Way, Suite 2, 4766,
Valley Cottage, NY 10989

www.scitusacademics.com

Notice

Printed in United States of America on Acid Free Paper

Preface

In groundwater exploration in hard rock media, it is very important to delineate structures such as geological contacts, faults and joints. Exploration of groundwater in hard rock terrain is a very challenging and difficult task when the promising groundwater zones are associated with fractured and fissured media. Since these structures control the groundwater flow and are more favorable sites for the groundwater accumulation. In this environment, the groundwater potentiality depends mainly on the thickness of the weathered/ fractured layer overlying the basement. The main aim of a geophysical investigation in a hard rock region is to solve different hydrogeological problems like depth of water table, basement rock, thickness of weathered zone/formation, structural and stratigraphic conditions relevant to groundwater conditions, and permeability of aquifers. The combination of various geophysical techniques with environmental conditions is essential to constrain the interpretation and reduce uncertainties.

Groundwater Geophysics in Hard Rock offers the methodology to groundwater exploration in hard rocks, several geophysical practices and combinations to be used, analysis of data with case studies and drilling results and the preparation of different utility maps.

Table of Contents

1

CHAPTER

Groundwater Exploration for Water Well Site Locations Using Geophysical Survey Methods

Haile Arefayne Shishaye[1*] and Semir Abdi[2]

[1]Department of Water Resource and Irrigation Engineering, Haramaya University, Ethiopia

[2]Hydrogeologist in Harar Regional Water Supply and Sewerage Bureau, Ethiopia

ABSTRACT

Groundwater exploration is the investigation of underground formations to understand the hydrologic cycle, know the groundwater quality, and identify the nature, number and type of aquifers. There are different groundwater exploration methods. Surface geophysical method is one of the groundwater investigation methods. One of the surface geophysical methods is therefore the vertical electrical sounding method. Vertical electrical sounding (VES) is one to provide valuable information regarding the vertical successions of subsurface geo-materials in terms of their individual thicknesses and corresponding resistivity values. It is rapid and much effective in estimating aquifer thickness of an area and is cost effective technique for groundwater study. The objective of this study was therefore to locate two well site locations using surface geophysical methods for water supply purposes. However, hydrogeological and geological investigations were also incorporated in addition to the geophysical surveying activities for the betterment of the project. Finally, the intended well site locations with their corresponding thickness and resistivity values were identified using the integrated approaches.

Keywords: Groundwater exploration; Geophysical survey; Water well; Haramaya University; Lake Haramaya watershed

INTRODUCTION

Water is an essence food and basic component of life. The need for water is strongly ascending and has a diversified function, which is not only important for drinking purposes but is also vital for any developmental activities. Nowadays, the use and sustainability of water is getting more complex due to population growth, urbanization and industrialization. Any development is related either directly or indirectly with water utilization.

For any developmental activity, both surface and groundwater sources are the main components depending on their quality and availability. In an area where surface water is not feasible for the desired activity, groundwater is the second alternative, if it has the anticipated amount and quality. Therefore, site investigation/ exploration, sometimes called a pre-construction evaluation, has to be performed primarily for an effective and sustainable utilization of groundwater resources.

Groundwater has become immensely important for the different water supply purposes in urban and rural areas of both the developed and developing countries [1]. However, groundwater exploration in hard rock terrain is a very challenging and difficult task, if the promising groundwater zones are associated with fractured and fissured media [2]. In such an environment, the groundwater potentiality depends mainly on the thickness of the weathered/ fractured layer overlying the basement.

Groundwater can be explored using different methods. The four major groundwater exploration methods are the areal method, surface method, subsurface method and esoteric methods. Among these methods, esoteric method is not based on science, mostly based on traditional indicators. Each of the above listed groundwater exploration methods have different sub-methods under them. Geophysical survey is therefore one of the sub-methods under the surface method of groundwater exploration [1]. This method is very important for both groundwater resource mapping and water quality evaluations [1]. Its application for groundwater exploration purposes has increased over the last few years due to the rapid advances in computer packages and associated numerical modeling solutions [2].

Geophysical survey incorporates the Vertical Electrical Sounding (VES) and Horizontal Profiling (HP) activities. The Vertical Electrical Sounding (VES) is currently being very popular with groundwater investigations due to its simplicity [3,4]. This geophysical survey method is the detection of the surface effects produced by the flow of electric current inside the earth [2,5]. It provides depth and thickness of various subsurface layers and their relative water yielding capacities [5].

The water supply for the Haramaya University community and the farmlands within the university campus comes from the wells drilled within the university compound. There are eight previously drilled wells within the university (Figure 1), excluding the newly drilled wells (new well 1 and 2, i.e., these two wells were drilled based on this study). However, the yield of the eight wells was not able to balance the gap between the demand of the university community and the current potential of the water supply.

Therefore, the objective of this study was to identify potential well site locations and to know the overall groundwater conditions of the study area using surface geophysical survey. More specifically, the study was aimed to understand the nature, number and type of aquifers within the study area so as to identify the potential well sites that could balance the water demand and supply.

MATERIALS AND METHODS

Description of study area

Haramaya University is located in the Lake Haramaya watershed. Lake Haramaya Watershed (Figure 2) is situated in the Eastern Ethiopia, found in Haramaya and partly in Kombolcha districts, Eastern Hararghe Zone, Oromia Regional State at 505 km East of Addis Abeba and 20 km north-west of Harar town. The university is therefore found in the Haramaya district. It lies at 9°22'03"- 9°27'12" North and 41°58'14"- 42°05'26" East.

The average annual rainfall in the watershed is 801 mm. The dry period (less than 30 mm per month) extends from October to January; while, the wettest month is August with an average rainfall of 144 mm. According to Shimelis [6], the annual average effective rainfall of Lake Haramaya watershed is about 673 mm.

Moreover, the annual mean temperature is 18°C. The months of October to January tend to be slightly warmer; while, June to September is the coolest period. The daily temperature in the study area ranges from around 10°C to 25°C.

Desk study

The study was conducted integrating both the desk study and field work activities. The desk study in this project involves evaluation of regional and sub-regional hydrogeological maps and evaluation of the previously developed Geological and hydrogeological maps of the watershed. Furthermore, it also involves investigations based on the enhanced Thematic Mapper satellite images at a resolution of 15 × 15 meters and evaluation of Geological and Hydrogeological map of Ethiopia (1:2,000,000).

Previous geological and hydrogeological work: Physiographically, the study area is part of the Harar Plateau areas, which is the upper part of Wabi-Shebele Basin (Figure 3). It is characterized by three major stratigraphic units including the Precambrian crystalline basement (mainly granite), Mesozoic sedimentary rocks (sandstone and limestone) and Quaternary sediments from old to young [7]. The

Precambrian basement complex includes the high grade metamorphic gneisses and migmatites, and the intrusive granites [8]. The sandstone rock unit outcrops on the eastern and western side overlying by limestone that is dipping towards west and underlined by the basement rocks (Figure 4). The quaternary sediment is covering the low lying area that comprises lacustrine sediments, alluvial and eluvial sediment (Figure 4).

Groundwater occurrence, circulation and storage properties of the area are determined greatly by the type of geology, geological contacts, geomorphology and rainfall patterns. The main aquifer system in the area is the unconsolidated material composed of loose coarse grained sand, pebbles and rock fragments, silt and clay materials [9], which is characterized by high-moderate to low productivities.

The basement (granite) unit has low permeability and productivity [7,8]; while, the sandstone and limestone units are characterized by moderate range of permeability and productivity [10]. The subsurface boundary condition of this porous, single-unconfined aquifer system is the replica of surface topography except on the North boundary [9].

Climate: Surface and subsurface water potential is dependent on various factors. These factors include the climate, geology, vegetation, topography, elevation and etc. Among the above listed factors, climate plays a great role in influencing meteorological parameters such as rainfall, temperature, relative humidity, evapotranspiration and wind speed, which are very important parameters for water resources evaluation.

Water is constantly evaporated from the earth and is precipitated back on the earth in the form of rain, snow, etc. Part of this precipitation infiltrates into the ground, raising the water table, and the major part flows as runoff; and the rest is lost in the form of evaporation and transpiration. The interaction between precipitation, infiltrated water, runoff and groundwater is rather complex. It is affected by physiographic, geologic, meteorologic and hydrologic factors as well as by human activities such as water use and pollution.

The study area belongs to the summer and spring rainfall region of the country. Rainfall comes through the Southwester lies. During the summer season (June, July and August) the sun is overhead north of the equator. The cyclone or low pressure cells are in southwest Asia. The anticyclones or high pressure cells are on the Atlantic and Indian Ocean around the Tropic of Capricorn. The Inter-Tropical Convergence Zone (ITCZ) is just north of Ethiopia or northeast of Ethiopia on the Red Sea and Gulf of Aden (Figure 5) and the area is under the influence of Atlantic equatorial westerlies and southern winds from the Indian Ocean. Both the Atlantic and Indian Oceans are sources of rainfall. The main source being the Atlantic as the southerlies from the Indian Ocean lose most of their moisture as rainfall over the Kenyan highlands. These southwester lies ascend over the western highlands of Ethiopia to produce the main rainy season; as a consequence the rainfall is mainly orographic. The summer rainfall has a wide coverage and all of the highlands and lowlands of the basin receive rain in varying degrees starting from June up to mid- September.

Aspects of groundwater recharge and discharge

Recharge: Groundwater recharge may be defined in a general sense as the downward flow of water reaching to the water table, and forming groundwater reservoir. Recharge may occur naturally from precipitation, rivers, canals, lakes, or as man induced phenomena (irrigation and urban recharge). Assessment of the mechanism and amount of recharge is fundamental for sustainable groundwater resources utilization and management. However, estimating the different recharge processes is not simple. It requires understanding the various processes that affect recharge and quantifying the spatial and temporal variability. Unfortunately, there is no direct means of measuring groundwater recharge at regional and sub-regional level.

Recharge estimation requires accounting the different factors. These include:

- Topography and geology
- Precipitation (magnitude, intensity, duration, spatial distribution)
- Runoff and ponding of water
- Irrigation (nature of irrigation scheduling, losses from canals and water courses, etc.)
- Rivers (rivers flowing into and leaving out of the area under consideration, rivers gaining water from or losing water to the aquifer, etc.).
- Soil zone (nature of the soil, depth, hydraulic parameters, variability of the soil spatially and with depth, rooting depth of the soil, and cracking of soil on drying out or swelling due to wetting)

- Unsaturated zone between soil and aquifer (flow mechanism through unsaturated zone, zones with different hydraulic conductivity, etc.)

- Ability of aquifer to store water and variation of aquifer condition with time

The recharge conditions for any area may comprise the above all types of recharge. However, the encountered types of recharge at the study area and its vicinity are:

Direct recharge - Water added to the groundwater reservoir in excess of soil moisture deficits and evapo-transpiration, by direct vertical percolation of precipitation through the unsaturated zone. In this case, the ultimate source of groundwater recharge in the site was found to be precipitation.

Indirect recharge - Percolation to the water table following runoff and localization in joints, as ponding in low lying areas and/or through the beds of surface water resources such as lakes and reservoirs.

Localized recharge - Resulting from horizontal surface concentrations of water in the absence of well-defined channels.

Discharge: Discharge of groundwater is manifested as springs, pumping, seepage zones and base flow of rivers. Groundwater discharge areas are often present in lowlands, local and regional depressions, and along the banks of rivers. Discharge areas are intimately linked with groundwater flow lines and the existence of geological structures. However, the main groundwater discharge in the study area was found to be pumping.

Aquifer characteristics: The aquifer characteristics of study areas are classified based on similar or nearly similar hydrogeological characteristics of geological structures. One of the basic elements of Hydrogeological study is production of hydrogeological map. The map has to include hydrogeological characteristics of hydro stratigraphic units, basically the geological characteristics of the area that includes degree of fracturing and degree of weathering, geological structures that favorite groundwater storage and movements and other parameters play significant role in designing capture and dewatering systems of groundwater. Besides schematizing the natural hydrogeological conditions, it is important to know the hydraulic/ hydrogeological parameters.

There are several important physical properties that govern the capability of an aquifer to store, transmit, and yield groundwater. Reliable interpretations and conclusions about the whole aquifer performance in an area are not possible without the accurate determination of these basic aquifer's physical properties.

According to Guesh et al. [9], the calculated transmissivity near existing wells in the study area was found ranging between 28.6-237.76 m2/day; while, values computed from pumping test result ranges between 41.67 m2/day and 336.9 m2/day. Moreover, the average hydraulic conductivity in the area was also found to be 4.85 m/day. Accordingly, the hydrogeologic nature of the aquifer in the study area was found to be favorable to drill water wells in.

FIELDWORK

In this study, the field work was conducted to investigate topography, local geological and hydrogeological conditions of the project area in relation to the developed maps. Moreover, geophysical investigations were also conducted using the SARIS 1000 Terrameter, which has digital LCD earth resistivity measuring instrument.

Hydrogeological and geophysical investigations: Groundwater potential evaluation for a given area/basin requires integrated approach. Detailed quantification of the amount of groundwater demand a detailed water balance study and defining the boundary conditions, establishment of the lateral and vertical extent of aquifers and confining beds. In the absence of detailed hydro meteorological and hydrogeological data, groundwater potential evaluation tends to be more semi-quantitative mainly based on geophysical investigation.

The method followed in this study, for the most part, was depending on the short-term field hydrogeological investigation and surface geophysical surveying. Both approaches provide information on the availability of groundwater in a semi-quantitative sense.

Geophysics provides no information on the exact amount of groundwater available in the subsurface. The amount can only be estimated when the geophysical survey is supported by the local hydrogeological features such as recharge potential, availability of permeable rocks, catchments areas, etc.

All the interpretations shown in the geophysical part of this study are made semi-quantitatively by integrating the hydrogeological field observations with the geophysical signature obtained from the VES data. The recommendations of the likely depth of drilling are made on the basis of the VES data and the nearby well data. The supervisor/ hydro geologist can recommend the accurate drilling depth during the drilling operation. For instance in some sites where geophysics does not give conclusive answers on the total depth of the aquifers, yield of the well can be defined by using provisional tests with the compressor of the drilling machine.

Data analysis and interpretation

Computer assisted iterative interpretation packages like IPI2 and Win RESIST software's of Schlumberger analysis were used to interpret the resistivity data gathered from the field. i.e., they were used to identify the thickness and depth of each hydro stratigraphic layer in the study area. Moreover, ArcGIS 10.2 was also used to locate the VES points and well locations within the study area.

RESULTS AND DISCUSSION

The Vertical Electrical Soundings were carried out at the selected areas. A total of eight VES points were surveyed. The UTM coordinates, i.e., Easting and Northing, of the sounding points were obtained with the help of a GARMIN II GPS receiver in Adindan datum.

The Vertical Electrical soundings were made using the following equipment:

- Wires and Reels

- Stainless steel electrodes
- Accessories such as alligator clips and hammer

The Schlumberger electrode configuration with maximum halfcurrent electrode separation (AB/2) of 100 to 150 m was used for the sounding survey. Current is injected into the ground using two current electrodes, A and B, placed at a distance AB/2 apart; and the potential drop that occurs between two other electrodes, M and N, placed near the center of the current electrodes is measured (Figure 6). The current electrode separation, AB/2, is progressively increased in steps so as to increase the depth of investigation, and at each step the measured current and potential readings are used to obtain the apparent resistivity of the ground.

Data processing and presentation

The VES is done by injecting a low current into the ground through stainless steel electrodes and then after the resistance of the earth material is measured during the passage of the current simultaneously. The VES data collected in the field was initially interpreted using IPI2 window based resistivity software just to get the layer parameters. The layer parameters so obtained have been used as starting models in an iterative least squares inversion program, and then the software automatically displays the layer resistivity, thickness and depth of the layers from the ground surface.

Interpretation results

Vertical Electrical Sounding (VES) Data and interpretation

Array: Schlumberger, Azimuth: E-W, Instrument: SARIS 1000, VES Location UTM E: 174362, N: 1041581, Gr. Elevation: 2027 m.a.s.l, VES I.D. No.: 1.

Site description: The VES was conducted at the flat land at Haramaya University within the Haramaya Well field. The VES was conducted to confirm the presence of geological structure, thereby to pinpoint appropriate site for drilling. On the bases of the geophysical data, a well can be drilled at this point to a maximum depth of 34-44 meters. This is because of the maximum depth of drilling in such a case can be extended up to ten plus the total depth of the layers found from the interpreting packages. The thickness of the productive aquifer in this case is 33.2 m (Figures 7 and 8).

Array: Schlumberger, Azimuth: E-W, Instrument: SARIS 1000, VES Location UTM E: 174130, N: 1041423 Gr. Elevation: 2023 m.a.s.l, VES I.D. No. 2.

Site description: VES was conducted at the flat land at the Haramaya University within the well field. The VES conducted to confirm the presence of geological structure, thereby to pinpoint appropriate site for drilling. The geophysical signature displays five resistive layer, the most promising aquifer zone is located at the fourth layer. However, the thickness of the aquifer is 18.9 m. So, on the bases of geophysical data, the well in this site can be drilled to a maximum depth of 23-33 meters (Figure 9).

Array: Schlumberger, Azimuth: E-W, Instrument: SARIS 1000, VES Location UTM E: 173923 N: 1041590, Gr. Elevation: 2022 m.a.s.l, VES I.D. No. 3.

Site description: VES was conducted at the flat land at the Haramaya University within the well field. The VES conducted to confirm the presence of geological structure, thereby to pinpoint appropriate site for drilling. The geophysical signature in this case displays three resistive layers; the most promising aquifer zone is located at the second layer. The thickness of the promising aquifer is around 43.5 m. On the bases of geophysical data, a well at this site can be drilled to a maximum depth of 45-55 meters (Figure 10).

Array: Schlumberger, Azimuth: E-W, Instrument: SARIS 1000, VES Location UTM E: 174612 N: 1041877, Gr. Elevation: 2027 m.a.s.l, VES I.D. No. 4.

Site description: VES was conducted at the flat land at the Haramaya University within the well field. The VES was conducted to confirm the presence of geological structure, thereby to pinpoint appropriate site for drilling. The geophysical signature displays three resistive layers; the most promising aquifer zone is located at the second layer. The thickness of the most promising layer is 52.3 m. Thus, on the bases of geophysical data, the well at this site can be drilled to a maximum depth of 54-64 meters. Accordingly, "New Well 2 (Figure 1)" was drilled at this site. The actual depth of the well, the location of the basement from the ground surface, was found to be 50 m (Figure 11).

Array: Schlumberger, Azimuth: E-W, Instrument: SARIS 1000, VES Location UTM E: 0174156 N: 1041783, Gr. Elevation: 2028 m.a.s.l, VES I.D. No. 5.

Site description: VES was conducted at the flat land at the Haramaya University within the well field. The VES conducted to confirm the presence of geological structure, thereby to pinpoint appropriate site for drilling. The geophysical signature displays four resistive layers; the most promising aquifer zone is located at the third layer. The thickness of the most promising layer is 37.2 m. Thus, on the bases of geophysical data, the well at this site can be drilled to a maximum depth of 40-50 meters (Figure 12).

Array: Schlumberger, Azimuth: N-S, Instrument: SARIS 1000, VES Location UTM E: 174975 N: 1042072, Gr. Elevation: 2025 m.a.s.l, VES I.D. No. 6.

Site description: VES was conducted at the flat land at the Haramaya University within the well field. The VES conducted to confirm the presence of geological structure, thereby to pinpoint appropriate site for drilling. The geophysical signature displays four resistive layers; the most promising aquifer zone is located at the third layer (Figure 13). The thickness of the most promising layer is 13.7 m. Thus, on the bases of geophysical data, the well at this site can be drilled to a maximum depth of 28-38 meters.

Array: Schlumberger, Azimuth: N-S, Instrument: SARIS 1000, VES Location UTM E: 173886 N: 1041482, Gr. Elevation: 2014 m.a.s.l, VES I.D. No. 7.

Site description: VES was conducted at the flat land at the Haramaya University within the well field. The VES conducted to confirm the presence of geological structure, thereby to pinpoint appropriate site for drilling. The geophysical signature displays five resistive layers; the most promising aquifer zone is located at the fourth layer. The thickness of the most promising layer is 31.5 m. Thus, on the bases of geophysical data, the well at this site can be drilled to a maximum depth of 51.1-61.1 meters. Accordingly, New well 1 (Figures 1 and 13) was drilled in this site. The actual depth of the well was found to be 52 m.

Array: Schlumberger, Azimuth: N-S, Instrument: SARIS 1000, VES Location UTM E: 173519 N: 1043268, Gr. Elevation: 2007 m.a.s.l, VES I.D. No. 8.

Site description: VES was conducted at the flat land at the Haramaya University within the well field. The VES conducted to confirm the presence of geological structure, thereby to pinpoint appropriate site for drilling. The geophysical signature displays four resistive layers; the most promising aquifer zone is located at the third layer. However, the thickness of the most promising layer is 11.07 m. Thus, on the bases of geophysical data, the well at this site can be drilled to a maximum depth of 13.5-23.5 meters (Figure 14).

CONCLUSION

The main goal of the study was to select potential well locations for water supply purposes. Among the several disciplines the preconstruction study, i.e., site investigation, which includes geological, Hydrogeological and Geophysical investigations need special attention that helps to minimize the project cost and increase quality and quantity of water.

Accordingly, the work has resulted in the identification of two shallow well development sites (VES-4 and VES-7), where new well 2 and new well 1 were drilled in, respectively. As indicated above, the work is mainly based on the geophysical investigations conducted in the study area, the conventional hydrogeological investigations and remote sensing (satellite imagery), and limited previous works in similar areas. The geophysical signature indicates that the recommended groundwater drilling depth is for shallow wells. For these wells the maximum depth of drilling may reach up to 60 m. However, as it was done in VES 4 and 7 (locations of new well 2 and new well 1), this figure may be changed based on the available water obtained during drilling and it is to be checked with provisional tests with the compressor. This is to be decided by the supervisor/field Hydro geologist.

REFERENCES

1. Al-garni MA (2009) Geophysical Investigations for Groundwater in a Complex Subsurface Terrain, Wadi Fatima, KSA: A Case History. Jordan Journal of Civil Engineering 3: 118-136.

2. Venkata Rao G, Kalpana P, Rao RS (2014) Groundwater Investigation Using Geophysical Methods-A Case Study of Pydibhimavaram Industrial Area. International Journal of Research in Engineering and Technology 3: 13-17.

3. Oyedele EAA, Olayinka AI (2012) Statistical Evaluation of Groundwater Potential of Ado-Ekiti, Southwest, Nigeria. Transnational Journal of Science and Technology 2: 110-127.

4. Ojoina OA (2014) Hydrogeophysical Investigation for Groundwater in Lokoja Metropolis, Kogi State, Central Nigeria.Journal of Geography and Geology 6: 1-10.

5. Jansen JR (2011) Geophysical Methods to Map Brackish and Saline Water in Aquifers. In Georgia Water Resources Conference. Florida: University of Georgia.

6. Shemelis G (2003) Performance Evaluation of Small Scale Irrigation Practices in Lake Alemaya Watershed and Methods of Improving Water Application Efficiency.

7. Kebede S (2013) Groundwater in Ethiopia, Features, Numbers and Opportunities. Springer Hydrogeology, Heidelberg, New York, Dordrecht, London.

8. Adem S (2012) Groundwater resource evaluation and sustainable management in Adelle and Haramaya dry Lake Catchment, eastern Ethiopia.

9. Zeru G, Alamirew T, Tadesse N, Olmana M (2015) Aquifer Characterization and Parameter Estimation of Haramaya Well Field from Surface Geophysical Method.Haramaya,Oromiya National Regional State.

10. Tesfaye C (1993) Hydrogeology of Ethiopia and water Resources Development, Ethiopian Institute of Geological Surveys, Ministry of Mines and Energy. Addis Ababa.

11. Haramaya University (2013) Participatory Integrated Development of Lake Haramaya Watershed, project proposal.

2

CHAPTER

A Multi-Scale Approach in Hydraulic Characterization of a Metamorphic Aquifer: What Can Be Inferred about the Groundwater Abstraction Possibilities

Antonella Baiocchi [1], Walter Dragoni [2], Francesca Lotti [1], Simone M. Piacentini [2] and Vincenzo Piscopo [1]

[1]*Department of Ecological and Biological Sciences, University of Tuscia, Viterbo 01100, Italy*

[2]*Department of Physics and Geology, University of Perugia, Perugia 06100, Italy*

ABSTRACT

: Hard-rock aquifers, which constitute a water supply source in many countries, are highly heterogeneous and defining a realistic model of an aquifer can be extremely complex. The objective of this study was to hydraulically characterize a metamorphic aquifer in a representative area of Italy and to identify the most appropriate approach for tapping of groundwater in this challenging environment. The results of surface fracture surveys, injection tests, pumping tests, and a simplified numerical model were compared. From the surface fracture survey, a model of the rock mass characterized by a well-developed discontinuity network and by a high frequency of discontinuities resulted. The injection tests showed the extreme heterogeneity and the lower hydraulic conductivity of the rock mass in comparison with the results of the pumping tests. The independent estimate of the hydraulic parameter resulting from numerical model highlighted a range of values higher than those resulting from the pumping tests. The study demonstrated that the continuum medium approach can be used in the case of hard-rock aquifers with a dense network of discontinuities. The multi-scale approach is recommended for investigating hydraulic heterogeneity and significantly helps to identify the most promising areas for well locations and their characteristics in relation to the style of fracturing.

Keywords: hard-rock aquifers; hydraulic properties; aquifer heterogeneity; numerical modeling; Italy

1. INTRODUCTION

Hard-rock aquifers constitute a water supply source in many regions of Africa, America, and Asia. These aquifers are generally characterized by a low yield due to their low intrinsic primary permeability and porosity. Several factors influence the occurrence of groundwater in hard rocks, such as lithology, geomorphology, tectonics, and climate conditions.

Many studies have shown that hard-rock aquifers are comprised of weathered mantle and fissured bedrock where the groundwater movement takes place. The hydraulic conductivity and storage of the weathered mantle and underlying fissured bedrock are derived from geodynamic and geomorphic processes. The fissures have been explained by several processes, such as lithostatic decompression, tectonic activity, cooling stress, and weathering processes. These two layers make up a composite aquifer followed in depth by the fresh basement, which is only locally permeable where tectonic fractures are present [1,2,3,4,5,6,7,8].

Hard-rock aquifers, such as those constituted by plutonic and metamorphic rocks, are recognized to be highly heterogeneous, particularly in relation to the different degrees of fracturing. This implies that the bedrock transmissivity can vary over many orders of magnitude [3,7,9,10,11,12,13]. Therefore, well yields are highly variable, and a high incidence of well failure occurs. Thus, defining a sustainable pumping rate is more difficult here than in other cases because of the extremely complex aquifer behavior. In this case, more so than in other cases, considerations of the discontinuous flow system, anisotropy, and heterogeneity are required to develop a reasonable model of the aquifer.

To plan for the optimal management and sustainable groundwater development of these aquifers when the bedrock is densely fractured, the continuum approach is the characterization method frequently used. This approach is less expensive than discrete-fracture or hybrid approaches. Even if several studies have concluded that the porous medium approximation is valid in densely fractured aquifers [14,15,16], the scientific debate is still open, particularly with regard to the scale at which the approximation becomes appropriate [17,18,19,20,21,22].

In Italy, where hard rocks outcrop over large areas in the Alps, Calabria, and Sardinia, the hydrogeological properties of plutonic and metamorphic rocks are not well known. The groundwater yields of these aquifers (generally less than $5 \times 10^{-3} m^3/s$ per km^2) are lower than those in the more common carbonate and alluvial aquifers (up to 3×10^{-2} m^3/s per km^2), which are widely used for the water supply. The interest in hard-rock aquifers is, thus, generally scarce; nevertheless, the large extent of these rocks and the scarcity of water resources in Sardinia (where hard-rock aquifers constitute 51% of the regional area) and Calabria regions (where hard-rock aquifers constitute 39% of the regional area) justify a thorough analysis [23]. Interest in these aquifers becomes even more relevant in the framework of the ongoing climate change: most of the scenarios regarding future climate scenarios in the Mediterranean area suggest a reduction in rainfall and, thus, a reduction in water resources [24,25].

This is the context of our study, which concerns a dam site in metamorphic rock in southern Sardinia. The dam site provides an opportunity to characterize a hydraulically fractured, metamorphic aquifer using a multi-scale approach. The results of the surface fracture surveys, injection tests, pumping tests, and simplified numerical model were compared to examine the heterogeneity of the hard-rock aquifer and to verify the degree of approximation of the porous medium approach. The objective of the study was to hydrogeologically characterize these low-permeability rocks and to examine which approach could better represent the fractured aquifer. This is a key point in the identification of the appropriate tapping method in a hydrogeological environment where water resources are generally scarce, but are fundamental to the drinking and irrigation water supply.

2. STUDY AREA

The area under examination is 25 km west of the town of Cagliari, in the Sardinia region (Figure 1), where the Paleozoic basement related to the Hercynian orogenic evolution outcrops. The Paleozoic basement consists of an originally-sedimentary succession (Lower Cambrian to Early Carboniferous) that was deformed during the Carboniferous period under the low metamorphic to anchizonal setting [26]. The Arburese Unit outcropping at the dam site consists of the Arenarie di San Vito Formation (Middle Cambrian to Lower Ordovician), which is several hundred meters thick and is composed of decimetric to metric alternations between micaceous metasandstones, quartzites, and metasiltstones. Clastic, poorly-cemented sediments, from gravels to sands and silts, of the Upper Pleistocene to the Holocene overlay the metamorphic basement [27]. The structural framework is mainly linked to the Hercynian Orogeny and Pliocene tectonic events. The Hercynian tectonics favored the thrust of the Arburese Unit over the Upper Ordovician to Lower Carboniferous succession and its deformation in large scale folding with axes of folds E–W and N–S oriented. The Pliocene tectonics shallow modified the structural setting through normal fault systems E–W, and N–S to NNW–SSE oriented [28,29].

The hydrogeological properties of the Paleozoic basement are not well known. It is only known that metamorphic rocks form a low-permeability aquifer (until 20 years ago, these rocks were considered to be impervious) where groundwater circulation occurs in the fissured zones close to the surface (on average, the first tens of meters). In the most heavily fractured zones (*i.e.*, near the faults), the groundwater flow is more active and can occur at a greater depth [30].

The study site corresponds to the building zone of the Medau Zirimilis Dam (Figure 1), a 50-m-high rockfill dam that was completed with an upstream impervious face and with a grout curtain 10–50 m deep. The drainage basin of the reservoir, 29.5 km^2 of extension, is characterized by a typical Mediterranean climate with mean annual precipitation between 700 and 900 mm and air temperature between 15 and 16 °C.

During the construction of the dam and the initial reservoir filling, several hydrogeological investigations were conducted. One of these was the measurement of the seepage through the rock foundation and laterally to the dam, that discharged in the drainage systems located at the base of the dam embankment and downstream of the grout curtain. These and other investigations highlighted the fact that the groundwater circulation involved the most fissured layers closest to the surface, including, on average, the first 50 m of the Paleozoic rocks, up to a depth of 100 m in the faulted zones. The most evident faulted zone has been found in the valley bottom at the site of the dam [30].

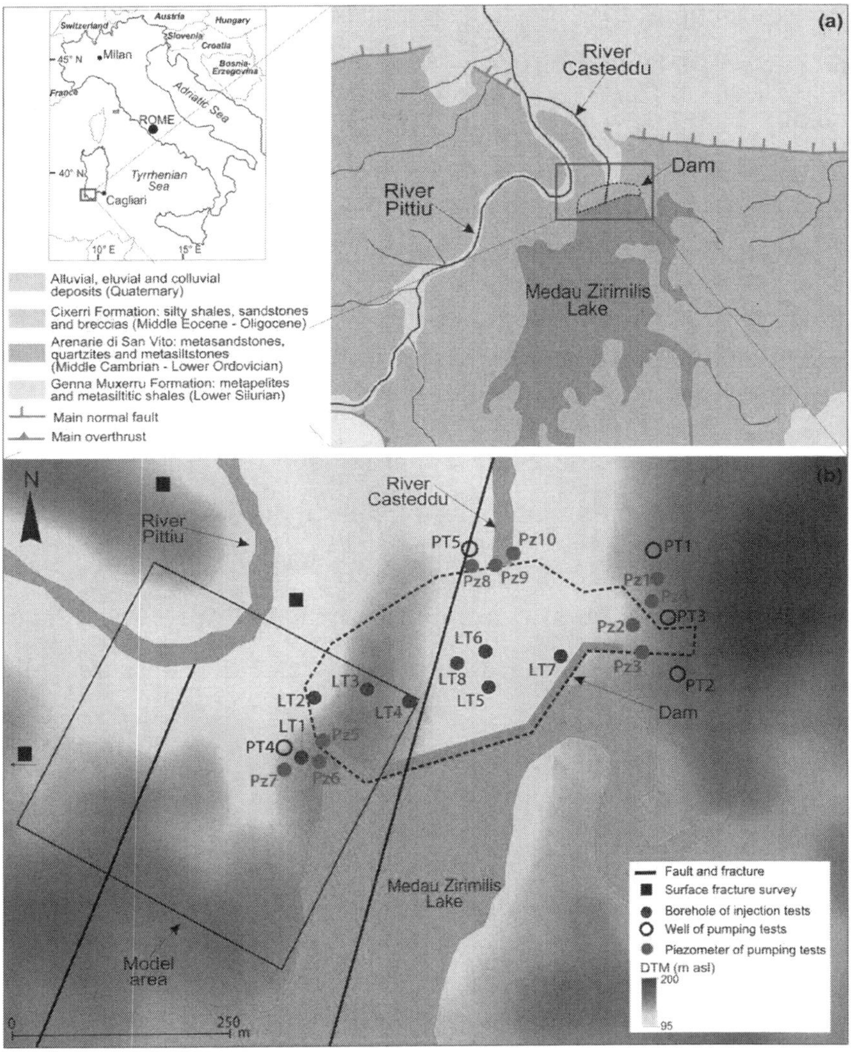

Figure 1. (**a**) Location of the study area; (**b**) Location of the surface fracture surveys, injection tests and pumping tests at the site of the Medau Zirimilis Dam.

3. INVESTIGATIONS, MATERIALS, AND METHODS

The results of the previous surveys [30] have been revised, integrated and appropriately processed to obtain the hydraulic characterization of the Palaeozoic rocks. The field investigations at the dam site included a surface fracture survey, injection tests and pumping tests (Figure 1).

The surface fracture survey concerned the Arenarie di San Vito Formation on three rock walls (ranging in size from 20 to 30 m^2) with different orientations located in the western sector of the dam (Figure 1): there were two vertical planes oriented to 60° N and 350° N and one horizontal plane. For each planar surface, the discontinuity orientation, the apparent aperture (measured by a feeler gauge with minimum determinable aperture of 0.05 mm), and the spacing between the discontinuities belonging to the same set were determined. The discontinuities considered were those having a trace length greater than 80% of the size of the measured plane (from 2 × 10 to 3 × 10 m).

The injection tests (Lugeon tests) concerned single-hole tests conducted on sealed sections with packers (from 3 to 5 m in length). Five pressure steps were applied (generally 0.2–0.5–1.0–0.5–0.2 MPa at the manometer), measuring the undisturbed water level before injection and making sure that the injected discharge was stabilized. A total of 76 tests were performed in eight boreholes penetrating the Arenarie di San Vito Formation (Figure 1) to a maximum depth of 60 m; one of the boreholes was drilled with a dip angle of 60° (LT8 in Figure 1).

During the initial reservoir filling, pumping tests were performed at five wells (Figure 1), with depths between 73 and 100 m (dip angle from 10° to 30°). The pumping tests lasted between 7.5 and 45 h at a constant rate between 2.1×10^{-3} and 3.5×10^{-3} m^3/s, with the observation of a drawdown in at least one piezometer nearby the tested well. The pumping data were interpreted using analytical techniques [31] and commercial software (Aquifer Test 4.1, Nova Metrix GM, Kitchener, Canada).

Based on the results of the hydraulic characterization of the site, a simplified numerical model was constructed for the western sector of the dam site (Figure 1), where one of the tested wells is located, and nearby the outcrops of the surface fracture surveys. The purpose of the numerical simulation was to examine the results of pumping tests to a further scale, different from that of the analytical solutions and of the other hydraulic tests. The model was implemented with the code MODFLOW-2000 (Groundwater Vistas 6 graphical user interface). The reconstruction of the DTM provided the top elevation. The initial heads were obtained using the interpolation of the heads measured during the initial reservoir filling.

4. RESULTS

The results are reported with reference to different methods of the hydraulic characterization and to the numerical simulation.

4.1. Discontinuity-Network Characterization

Two discontinuity types were identified in the examined outcrops. The first type consisted of regular bedding planes that were continuous at the scale of the outcrops and delimited the various metasandstone and metasiltstone beds. The second type of discontinuity consisted of sub-vertical straight joints of varying orientations that cut the bedding planes.

The projection of the discontinuity poles on a stereographic net [32,33] made it possible to group the systematic discontinuities into three different sets (Figure 2). Table 1 lists the average values of the dip direction, dip angle and Fisher dispersion coefficient of each set.

The measured spacing approaches a log-normal distribution for the three sets. The apertures were measured on discontinuity segments when they were not altered by the weathering phenomena and ranged from <0.05 mm (instrumental measurement limit) to 1 mm. None of the different distributions tested (normal, log-normal, logistic, gamma, and Weibull) perfectly represented the measured data. The log-normal distribution best represented the majority of the data, even if it contained a greater dispersion for the minimum and maximum values of the parameter. The statistics regarding the discontinuity spacing and the apparent aperture are listed in Table 1.

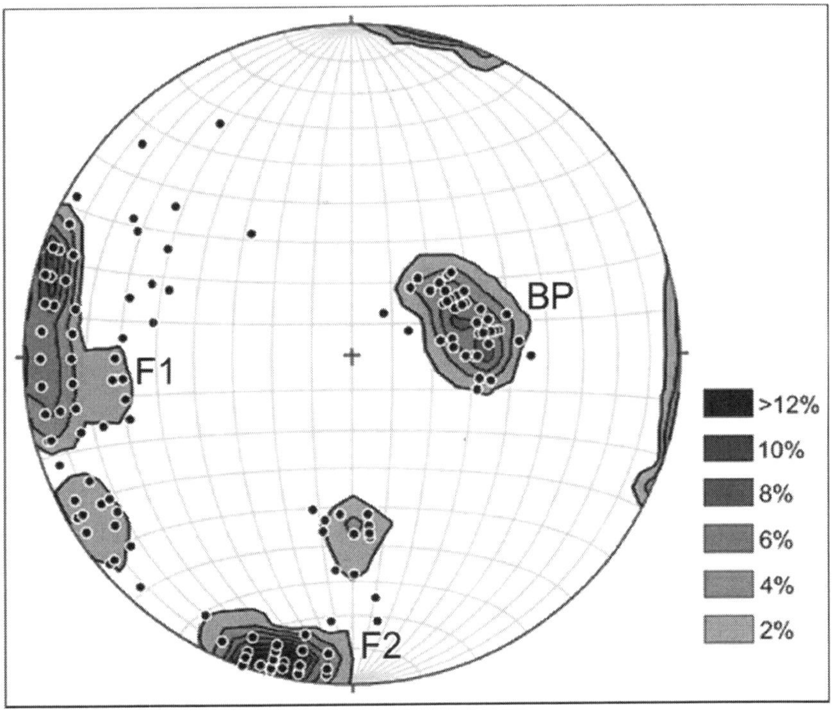

Figure 2. Equal-area projection of discontinuity poles measured in three outcrops (contour interval = 2%) showing the bedding plane (BP) and two fracture sets (F1 and F2).

A simplified model of the fracture network has been considered for a first estimate of the hydraulic properties of the rock mass.

If the fractured rock mass is assumed to have an impermeable matrix, if the discontinuities are considered as parallel plate model, with an aperture a and a spacing d, and if the groundwater moves by laminar flow in the discontinuities, the hydraulic conductivity k_i of each set can be estimated according to the cubic law [34]:

Table 1. Characteristics of the three distinguished discontinuity sets, bedding plane (BP) and two sub-vertical fracture sets (F1 and F2).

Characteristic	Statistics	Set 1: BP	Set 2: F1	Set 3: F2
Orientation	Number of measurements	48	96	68
	Mean dip direction (°)	255.1	91.9	11.9
	Mean dip angle (°)	29.2	78.1	75.8
	Fisher dispersion coefficient	68.7	13.0	21.6
Spacing	Number of measurements	64	74	133
	Mean (cm)	26.20	24.45	10.47
	Standard deviation (cm)	1.77	2.01	1.38
Aperture	Number of measurements	97	187	161
	Mean aperture (mm)	0.12	0.15	0.15
	Standard deviation (mm)	0.19	0.26	0.21

$$k_i = \frac{a^3 g}{12 \upsilon d} \quad (1)$$

where g is the gravitational acceleration and υ is the kinematic viscosity of water.

The values of a and d introduced in Equation (1) are the mean values of the apparent aperture and spacing resulting from the statistical analysis (Table 1). The calculated hydraulic conductivities of the different sets ranged from 5×10^{-6} to 3×10^{-5} m/s (Table 2).

Table 2. Estimated hydraulic conductivity of each discontinuity set, bedding plane (BP) and two sub-vertical fracture sets (F1 and F2).

Set	k (m/s × 10^{-5})
Set 1: BP	0.54
Set 2: F1	1.13
Set 3: F2	2.64

If the discontinuities are well-connected and if the flows between the discontinuity sets do not interfere, the hydraulic conductivity tensor of the fractured rock mass [*K*] can be calculated based on the orientation of the discontinuity sets [35]:

$$[K] = \sum_{i=1}^{n} k_i \times H_{jl}^{i} \; j,l = x,y,z$$
$$H_{jl}^{i} = [I - n^T \times n] \qquad (2)$$

where k_i is the hydraulic conductivity of the *ith*-discontinuity set; *I* is the unit matrix; and *n* is the direction cosine vector whose components are expressed in terms of the dip azimuth and dip angle of the discontinuity set (with the following coordinate system: *y*-axis oriented north, *x*-axis oriented east and *z*-axis oriented along the vertical).

The principal axes of anisotropy and the diagonal components of [*K*] in these directions were determined using the eigenvalues and eigenvectors of the matrix [*K*]. The eigenvalues and the orientation of the principal axes of the tensor are listed in Table 3.

Table 3. Values and orientation of the major, intermediate and minor components of the hydraulic conductivity of rock mass.

Components	Value (m/s × 10^{-5})	Trend (°)	Plunge (°)
Minor	1.58	201	15
Intermediate	3.19	109	6
Major	3.84	359	74

4.2. Injection Tests

The injection tests processing first included a study of the relation between the flow rate (q) and the net injection pressure (H_0). In some cases, the relations highlighted the opening and erosion processes in the discontinuities, while, in other cases, the void filling and essential laminar flow were noted.

The hydraulic conductivity k_L in each test interval was calculated by applying the following equation [36]:

$$k_L = \frac{q}{lH_0}\left[\frac{1}{2\pi}\left(1 + ln\frac{l}{2r}\right)\right] \quad (3)$$

where l and r are the length of the injection section and the borehole radius, respectively; and q and H_0 are the flow rate and the net injection pressure values under laminar flow, respectively.

The calculated values of k_L fell within three orders of magnitude (10^{-8}–10^{-6} m/s), with a geometric mean of 2.81×10^{-7}m/s. Figure 3 shows the k_L values of each borehole plotted *versus* the average depth of the injection section. At depths greater than 25 m, there was a significant decrease in the hydraulic conductivity, with the exception of the inclined borehole (LT8 in Figure 3). Above a depth of 25 m, the boreholes that were located in the valley bottom (LT4-LT8 in Figure 3), corresponding to a faulted zone, exhibited a higher k_L.

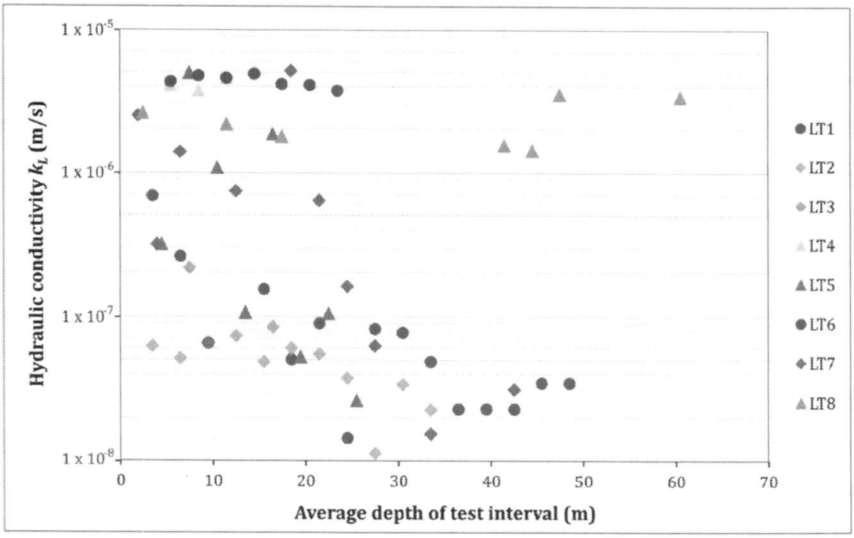

Figure 3. Hydraulic conductivity k_L from the injection tests *versus* the average depth of the injection section.

4.3. Pumping Tests

The pumping test data diagnosis included a comparison of the drawdown plots of each piezometer with theoretical models to determine the aquifer parameters. This analysis showed that the Theis [37], leaky aquifer [38] and double porosity [39] models could explain most of the measured drawdowns; in some cases, two or three theoretical models interpreted the same drawdown-time trend without significant differences in the calculated aquifer parameters, *i.e.*, the transmissivity and storage coefficient. For wells with more than one piezometer, there was a simultaneous response in the pumping, and generally, a similar shape for the drawdown curve for the different observation wells was reported, as observed in Figure 4.

Table 4 summarizes the results of the aquifer parameters, which were calculated as an average value when more than one model presented a good fit with the measured data and with the statistics of the curve matching. The overall transmissivity and storage coefficients varied from 2.3×10^{-4} to 3.6×10^{-3} m^2/s and from 3.1×10^{-5} to 2.6×10^{-2}, respectively. The transmissivities were converted into hydraulic conductivities (K_P) based on the saturated aquifer thickness intercepted by the tested well. The values of K_P fell in the range of 4.6×10^{-6}–4.1×10^{-5} m/s, with a geometric mean of 1.68×10^{-5} m/s.

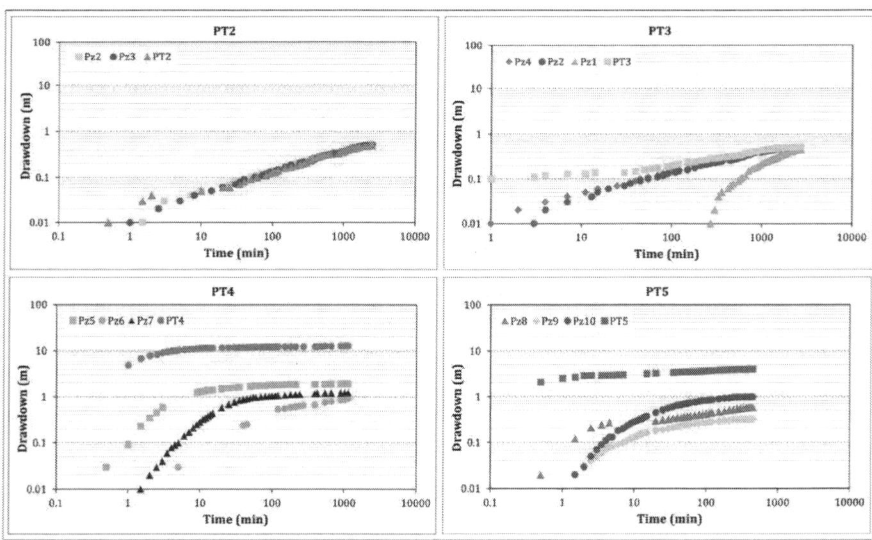

Figure 4. Drawdown-time plots resulting from the pumping tests for the wells with more observation piezometers.

Table 4. Results of the pumping tests: *Q*, constant rate of pumping; H_{sat}, saturated thickness of the aquifer intercepted by the well; *T*, transmissivity; K_P, hydraulic conductivity; *S*, storage coefficient; RSS, sum of squared residuals.

Well	Q (m³/s)	H_{sat} (m)	Piezometer	T (m²/s)	K_P (m/s)	S	RSS
PT1	2.1×10^{-3}	50	Pz1	4.89×10^{-4}	9.78×10^{-6}	2.61×10^{-2}	0.001
PT2	2.3×10^{-3}	41	Pz2	1.44×10^{-3}	3.51×10^{-5}	1.85×10^{-3}	0.024
			Pz3	1.42×10^{-3}	3.46×10^{-5}	5.11×10^{-3}	0.024
PT3	2.4×10^{-3}	53	Pz4	1.51×10^{-3}	2.85×10^{-5}	7.29×10^{-3}	0.021
			Pz2	1.41×10^{-3}	2.66×10^{-5}	4.56×10^{-3}	0.023
			Pz1	4.46×10^{-4}	8.41×10^{-6}	1.80×10^{-2}	0.002
PT4	2.5×10^{-3}	50	Pz5	2.32×10^{-4}	4.64×10^{-6}	3.15×10^{-5}	0.045
			Pz6	9.50×10^{-4}	1.90×10^{-5}	6.26×10^{-4}	0.013
			Pz7	2.33×10^{-4}	4.66×10^{-6}	5.22×10^{-4}	0.035
PT5	3.5×10^{-3}	86	Pz8	3.57×10^{-3}	4.15×10^{-5}	9.92×10^{-4}	0.023
			Pz9	3.44×10^{-3}	4.00×10^{-5}	8.83×10^{-4}	0.001
			Pz10	8.24×10^{-4}	9.58×10^{-6}	2.64×10^{-4}	0.002

4.4. Groundwater Flow Model

The hydrogeological characterization of the site was completed by a potentiometric surface reconstruction over the entire dam site using the head data measured in 70 wells and boreholes during the initial reservoir filling when the artificial lake level was at 121 m a.s.l. [30]. The head contour line deformation due to the lake and due to the gaining river was obtained by adding head points along the boundary conditions; the dam effect (*i.e.*, the upstream impervious face and grout curtain) was obtained by applying an impervious bound in the Kriging interpolation technique. From Figure 5, where the head distribution and the standard error ranges associated with the Kriging prediction are shown, it appears that the head contour lines generally replicate the topography. On the left of the dam, the lake feeds groundwater which discharges into River Pittiu and laterally bypasses the grout curtain flowing towards the rockfill embankment of the dam.

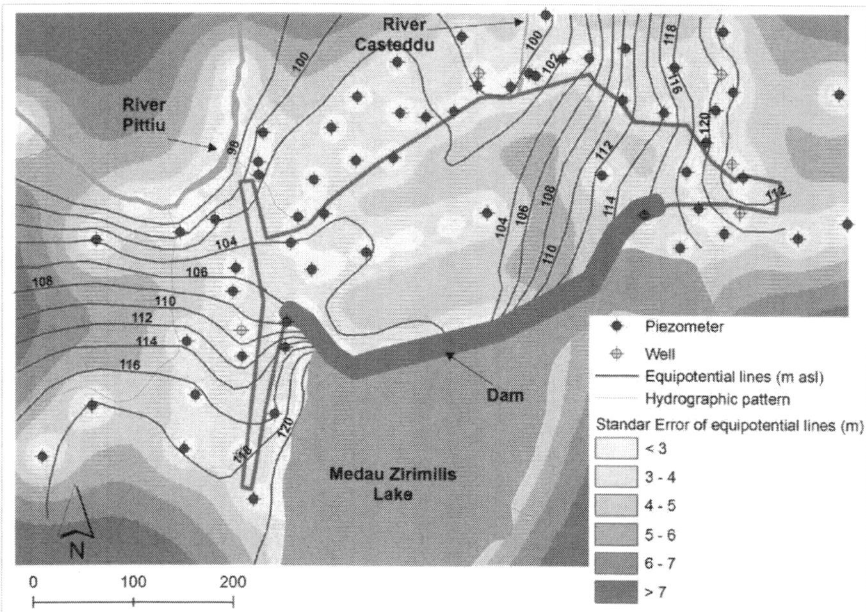

Figure 5. Potentiometric map at the site of the Medau Zirimilis Dam reconstructed by the kriging interpolation technique and the associated ranges of standard error when the artificial lake level was at 121 m a.s.l.

Thanks to the presence of physical boundary conditions (lake and river) and to the presence of both pumping tests and surface fracture survey, the western sector of the dam was selected for the numerical simulation (Figure 1).

The model involved an area of 86,750 m^2 (Figure 6). The grid consisted of one layer of 8736 cells with a telescopic refinement around the pumping well (PT4) and side dimensions from 0.33 to 8 m. As the investigations revealed, the bottom of the aquifer was not well defined; the available stratigraphic logs where compared with the results of injection and pumping tests and the model bottom was set at a depth of 50 m from ground surface. The aquifer was considered continuous, unconfined and isotropic, as resulted from field investigations.

The boundary conditions included: constant head boundary for the lake (*i.e.*, water level at 121 m a.s.l.); drain boundary condition for the gaining stream (*i.e.*, River Pittiu) and for section of the dam embankment where flow was measured in the drainage systems; no-flow cells in sectors outside the domain of interest or along groundwater divides (western side); well boundary condition for the pumping well (*i.e.*, PT4 well); constant flux boundary condition for the southern boundary (Figure 6). No precipitation event occurred in the days before and during the pumping tests, therefore no areal recharge was applied.

The days in which the pumping test was performed, a constant pumping rate of 2.5×10^{-3} m^3/s from the well PT4 was measured. The same day the lake level was 121 m a.s.l. and a flow of 6.7×10^{-3} m^3/s was measured from the horizontal drain located at the base of the embankment in the left sector of the dam [30]. If we consider the portion of the drain that falls within the modeled area, a maximum outflow of 3×10^{-3} m^3/s is expected from this model boundary. Outflow from River Pittiu was not measured during the days of the test.

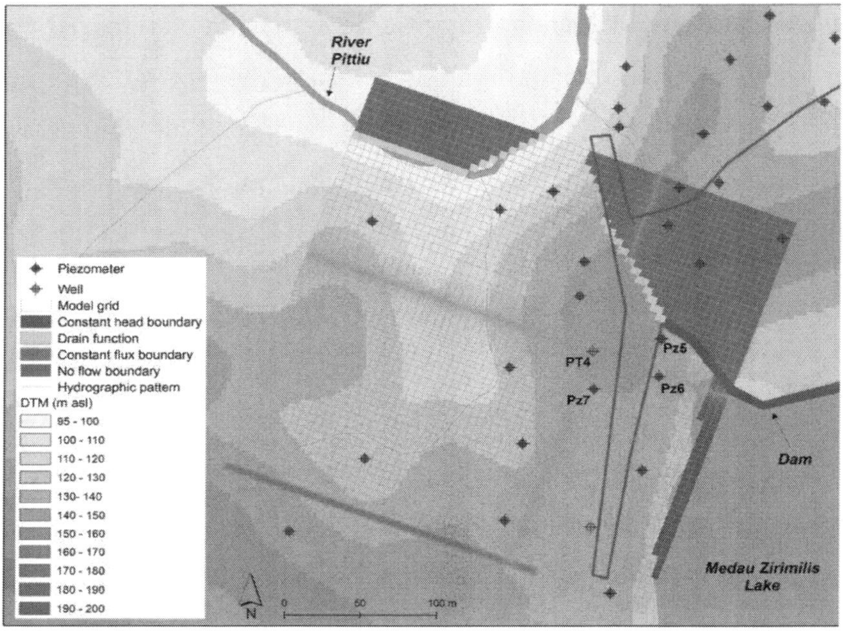

Figure 6. Grids and boundary conditions of the numerical model. DTM: digital terrain model.

Inflows from the southern and lake boundaries were estimated. For the southern boundary, considering the surface area of the upstream potential groundwater basin drained from this section, an area of about 2 km^2 is obtained. Base flow measures are available for an instrumented basin which is morphologically and geologically comparable with the basin under investigation, resulting between 2×10^{-3} and 4×10^{-3} m^3/s per km^2 (*i.e.*, River Fluminimaggiore; [40]).

According to this information, an inflow from 4×10^{-3} to 8×10^{-3} m^3/s is expected from the southern boundary. A second estimate considers the hydraulic gradient and transmissivity of this sector of the dam site and the length of boundary, obtaining an inflow ranging between 5×10^{-3} and 6×10^{-3} m^3/s. Using this last approach for the lake boundary, an inflow ranging between 4×10^{-3} and 5×10^{-3} m^3/s results.

The model underwent an initial calibration through the trial-and-error technique. The main purposes of this step were to test the numerical stability of the model and to roughly reduce the initial difference between the observed and calculated values of the observation points. This step preceded the application of the parameter estimation code PEST 13.0 [41], which was applied in two different modes: classical zone calibration and pilot point calibration [42,43].

A steady-state calibration was used as the first time period of the transient simulation. A second time period, divided into 20 steps, was introduced with pumping from PT4 at a constant rate (2.5×10^{-3} m^3/s) for 19 hours. Transient data were calibrated for the pumping well (PT4) and Pz5 and Pz7 piezometers.

A sensitivity analysis of the transient model highlighted the significant influence of the bottom elevation, which was only known in coincidence of boreholes and piezometers. Therefore the results were expressed by translating the hydraulic conductivity into transmissivity.

In Figure 7, simulated contour lines at the end of 19 h and transmissivity distribution are reported. In the same figure the simulated and observed heads of all wells and piezometers are shown, as well as the trends of drawdown in the PT4 well and in the Pz5 and Pz7 piezometers. The calibrated values of the specific yield are within the order of magnitude of 10^{-4}. The water balance is shown in Table 5 in comparison with measurements and analytical computations of the flows.

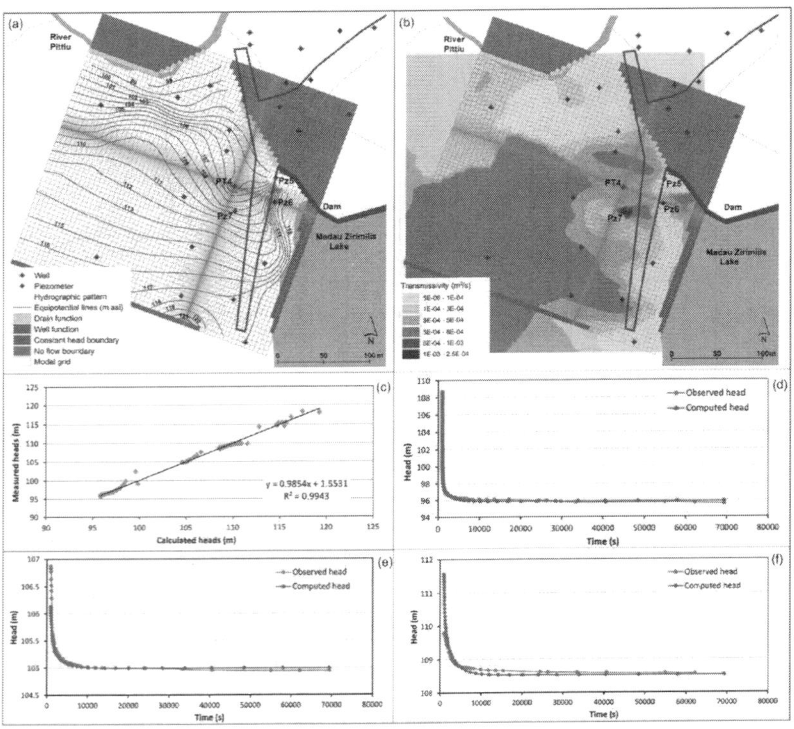

Figure 7. Results of the transient-state calibrated model showing (**a**) the potentiometric contour map; (**b**) the transmissivity distribution; (**c**) the scatter plot of the measured and calculated heads; (**d**) the comparison of the observed and calculated heads for the PT4 tested well; (**e**) for the Pz5 piezometer and (**f**) for the Pz7 piezometer.

Table 5. Groundwater budget resulting from the numerical model compared with the estimated terms from the conceptual model.

Model	Inflow (× 10^{-3} m^3/s)			Outflow (× 10^{-3} m^3/s)			% Error
	Lateral Boundaries	Lake	Storage	Well	River	Lateral Boundaries	(in-out)
Numerical Model	6.84	3.10	0.08	2.50	5.69	1.81	0.2
Conceptual model	4–8	4–5	-	2.50	na	3	-

Notes: na: not available.

Composite sensitivities associated with pilot points were calculated for the transient model with reference to the hydraulic conductivity, *i.e.*, the parameter that mainly affects the results. Highest values of composite sensitivity are associated to the pilot points corresponding with the pumping well PT4 and those falling in its surroundings (Figure 8), where the highest number of observations is located. Therefore, we can assume a good representativeness of the calibrated hydraulic conductivity in this zone, where the calibrated parameters give ranges of transmissivity from 5.0×10^{-6} to 2.5×10^{-3} m^2/s.

5. DISCUSSION

Investigations concerning fractured metamorphic rocks in a representative area of Sardinia can be used to hydrogeologically characterize these low-permeability rocks and to examine which approach can be used to represent the fractured aquifer. By interfacing the results of surface fracture surveys, injection tests and pumping tests with those of a flow numerical model, we inspected if the porous equivalent approach could (or could not) provide a good approximation of a fractured aquifer.

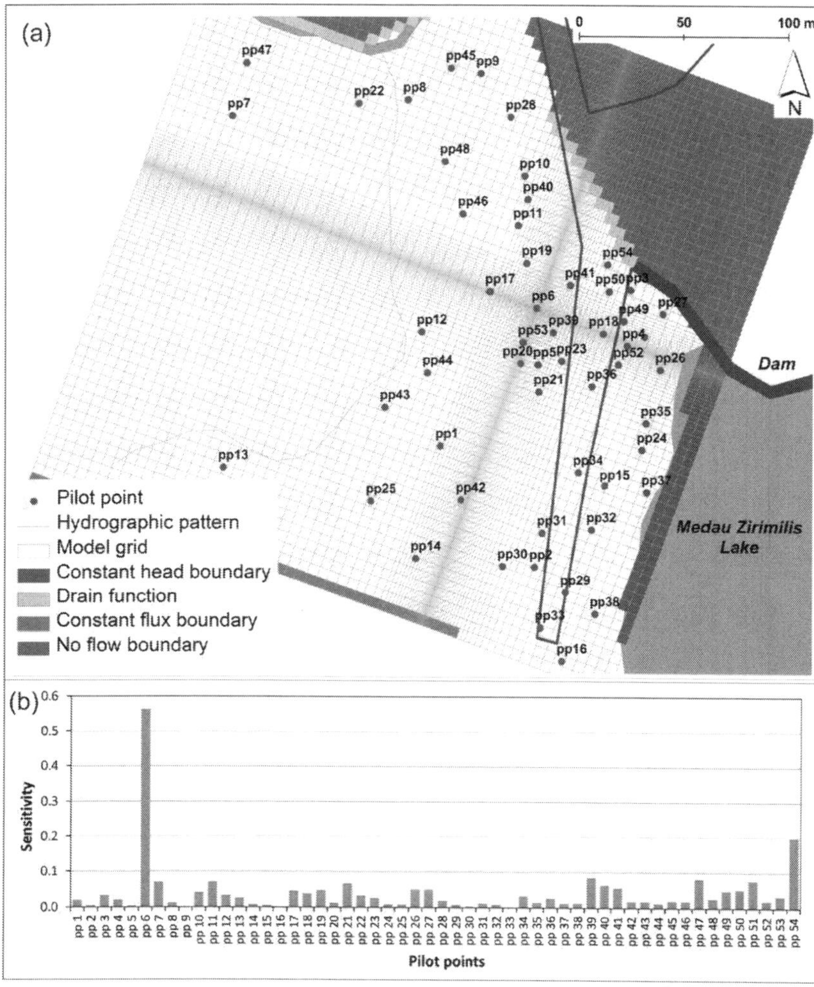

Figure 8. (**a**) Distribution of the pilot points in the model domain and (**b**) values of composite sensitivities for each point.

The surface fracture survey provides a model of rock mass that is characterized by a well-developed discontinuity network (three discontinuity sets), a high frequency of discontinuities (spacing of less than 0.5 m) and comparable values of the average apparent aperture of the different sets (approximately 0.1 mm). Therefore the results of the surface fracture survey implied the possibility of considering the fractured medium as an equivalent porous medium at the outcrop scale, because the discontinuity network gives rise to a reduced representative elementary volume (less than 1 m^3) for the rock mass.

The discontinuity network produces a slight anisotropy in the medium, with the major principal component of the hydraulic conductivity directed vertically downward, as expected given the predominance of vertical discontinuities over horizontal ones in the studied area. However, the modules of the hydraulic conductivity determined for the rock mass should be considered in relative terms and not in absolute terms because the apparent aperture does not correspond to the hydraulic aperture. A greater reliability can be assigned to the orientation of the three principal components, independent from the aperture.

The injection tests clearly showed the extreme heterogeneity of the rock mass, as highlighted by the wide range of values determined for the hydraulic conductivity (10^{-8}–10^{-6} m/s). This can be interpreted by considering the variability of the rock mass fracturing in relation to the proximity to the faulted zone in the valley of the dam site. The highest values of hydraulic conductivity were found for the inclined borehole, which was close to the fault zone and, thus, intercepted a great number of vertical fractures. With the exception of the inclined borehole close to fault zone, the hydraulic conductivity was observed to diminish below a depth of 25 m.

The hydraulic conductivity derived from the pumping tests covered a lower range of values (10^{-6}–10^{-5} m/s) than did those calculated using the injection tests. No particular trend was evident from the comparison of the values of hydraulic conductivity obtained for piezometers located in different directions from the pumped well. The variation in the parameter is therefore attributable to the heterogeneity of the aquifer rather than to its horizontal anisotropy. Moreover, if we consider the simultaneous response to pumping and the similar shape of the drawdown curves for the different piezometers when the pumping was performed from the same well, the porous medium approximation appears appropriate.

The comparison between the hydraulic conductivity determined by the injection tests and the pumping tests revealed values that were substantially lower in the first case (Figure 9). This can be attributed to the limited volume of rock mass affected by the injection tests compared to that affected by the pumping tests. In both cases, the hydraulic heterogeneity of the rock mass was evident. The injection tests and the pumping tests also converged and indicated higher hydraulic conductivities near the fault that runs in the valley of the dam site (Figure 3 and Table 4). Instead, the limit for the determination of the aperture by the surface survey did not permit a direct comparison of the hydraulic conductivity values determined using this method. However, if the average hydraulic conductivity (*i.e.*, the geometric mean) calculated from the pumping tests and the average spacing of the three sets determined from the surface survey are introduced in Equation (1), the average aperture value would be 0.16 mm, which is comparable with those measured in outcrop. However, the range of hydraulic conductivity resulting from the surface survey (10^{-6}–10^{-5} m/s) is consistent with that evaluated by pumping tests (Figure 9).

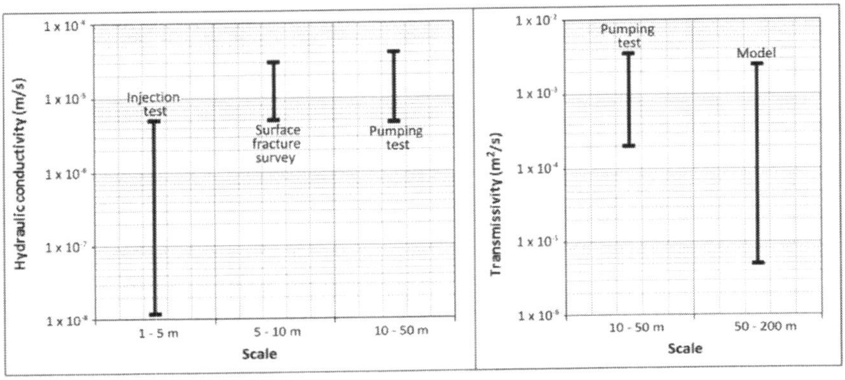

Figure 9. Hydraulic parameters and scale of measurement of the different methods used.

This scale effect of the parameter is expected in heterogeneous media [44,45,46,47] because the volume investigated by the injection tests was significantly smaller than that of the pumping tests. In the graph of Figure 9 one can see that the range of hydraulic conductivity values obtained from the surface fracture survey is consistent with the increase of hydraulic conductivity with the increase of the measuring scale. This implies that the results of surface fracture survey can be considered in the characterization of hard rocks when the discontinuity apertures are measured on the unweathered portions of discontinuity, at least to identify the order of magnitude of the parameter.

The numerical model considered one representative pumping test and provided additional information on the hydrogeological characterization of the hard rocks.

The results of the field investigations together with the model output would encourage the equivalent porous approach, at least in geological situations similar to that here considered. The estimate of transmissivity resulting from the model highlighted a range of values higher than those resulting from the pumping tests. However, the transmissivity values determined in the pumping tests (from 10^{-4} to 10^{-3} m^2/s) fell in the range of the calibrated values of the model (from 10^{-6} to 10^{-3} m^2/s), with a common upper limit of the parameter (Figure 9). This still seems to be consistent with what is reported in the literature, *i.e.*, that the increase in average permeability apparently does not continue from the pumping test to a regional scale [44,45]. On the other hand, four orders of magnitude of transmissivity obtained from the model compared with two orders of magnitude from the pumping tests, highlight a high hydraulic heterogeneity of the aquifer, this time in agreement with the results of the injection tests.

6. CONCLUSIONS

The multi-scale characterization of the metamorphic rocks of the area studied herein demonstrated that the continuum medium approach can be used when a dense network of discontinuities exists (representative elementary volume of few cubic meters), as in the case of the fractured aquifer examined. What strongly affects this approach is the hydraulic heterogeneity of the medium. The multi-scale approach is thus recommended for investigating the hydraulic heterogeneity, and it can help to manage uncertainty by placing greater trust in the obtained results. In particular, it is evident that the surface fracture survey enhances the hydraulic characterization of fractured rocks, when the fracturing affects the layers closest to the surface, as the general case of the hard-rock aquifers.

Regarding the transmissivity of the examined hard-rock aquifer, it falls in the classes III and IV of the classification proposed by Krasny [48], corresponding to rocks from low to intermediate transmissivity magnitude. This aquifer, commonly believed to be not productive, may be of interest for the local water supply, especially if one considers the large extent of these rocks and the scarcity of water resources in Sardinia. The same concept could be applied in other regions of the world where these aquifers outcrop over very large areas. In this regard, our findings confirm that the most promising areas for well location correspond to valleys close to faulted zones where transmissivity increases; in these specific zones a relative higher yield of wells is expected, especially in case of boreholes inclined according to an angle which can be determined after the fracture characterization.

ACKNOWLEDGMENTS

The authors wish to thank the two anonymous reviewers for their constructive comments and suggestions. Vincenzo Piscopo want to remember recently deceased Pietro Celico to have introduced him with passion and dedication to hydrogeology.

AUTHOR CONTRIBUTIONS

All of the authors contributed extensively to the work. Vincenzo Piscopo led the research. Antonella Baiocchi, Francesca Lotti and Simone Piacentini performed the data analysis and numerical modeling. Walter Dragoni contributed to the manuscript review. All authors contributed to the manuscript preparation, discussing and sharing the conclusions.

CONFLICTS OF INTEREST

The authors declare no conflict of interest.

REFERENCES

1. Banks, D.; Banks, S. Hydrogeology of Hard Rocks. In Proceedings of the Memoires of the 24th Congress, International Association of Hydrogeologists, Oslo, Norway, 28th June–2nd July 1993; volume 28, pp. 1–684.

2. Stober, I.; Bucher, K. *Hydrogeology of Crystalline Rocks*; Kluwer Academic Publishers: Dordrecht, The Netherlands, 2003.

3. Taylor, R.; Howard, K. A tectono-geomorphic model of the hydrogeology of deeply weathered crystalline rock: Evidence from Uganda. *Hydrogeol. J.* **2000**, *8*, 279–294.

4. Cho, M.; Ha, K.-M.; Choi, Y.-S.; Kee, W.-S.; Lachassagne, P.; Wyns, R. Relationship between the Permeability of Hard-Rock Aquifers and Their Weathered Cover Based on Geological and Hydrogeological Observation in South Korea. In Proceedings of the IAH Conference on

"Groundwater in fractured rocks", Prague, Czech, 15–19 September 2003.

5. Dewandel, B.; Lachassagne, P.; Wyns, R.; Maréchal, J.C.; Krishnamurthy, N.S. A generalized 3-D geological and hydrogeological conceptual model of granite aquifers controller by single or multiphase weathering. *J. Hydrol.* **2006**,*320*, 260–284.

6. Krasny, J.; Sharp, J.M. *Groundwater in Fractured Rocks*; Taylor & Francis: London, UK, 2007.

7. Foster, S. Hard-rock aquifers in tropical regions: Using science to inform development and management policy.*Hydrogeol. J.* **2012**, *20*, 659–672.

8. Sharp, J.M. *Fractured Rock Hydrogeology*; Taylor & Francis: London, UK, 2014.

9. Uhl, V.W.; Sharma, G.K. Results of pumping tests in crystalline-rock aquifers. *Ground Water* **1978**, *16*, 192–203.

10. Howard, K.W.K.; Hughes, M.; Charlesworth, D.L.; Ngobi, G. Hydrogeologic evaluation of fracture permeability in crystalline basement aquifers of Uganda. *Hydrogeol. J.* **1992**, *1*, 55–65.

11. Chilton, P.J.; Foster, S.S.D. Hydrogeological characteristics and water-supply potential of basement aquifers in tropical Africa. *Hydrogeol. J.* **1995**, *3*, 3–49.

12. Maréchal, J.C.; Dewandel, B.; Subrahmanyam, K. Use of hydraulic tests at different scales to characterize fracture network properties in the weathered-fractured layer of a hard rock aquifer. *Water Resour. Res.* **2004**, *40*.

13. Banks, D.; Gundersen, P.; Gustafson, G.; Makela, J.; Morland, G. Regional similarities in the distributions of well yield from crystalline rocks in Fennoscandia. *Norges Geol. Unders. Bull.* **2010**, *450*, 33–47.

14. Lee, C.H.; Chang, J.L.; Deng, B.W. A continuum approach for estimating permeability in naturally fractured rocks.*Eng. Geol.* **1995**, *39*, 71–85.

15. Scanlon, B.R.; Mace, R.E.; Barrett, M.E.; Smith, B. Can we simulate regional groundwater flow in a karst system using equivalent porous media models? Case study, Barton Springs Edwards aquifer, USA. *J. Hydrol.* **2003**, *276*, 135–158.

16. Lemieux, J-M.; Therrien, R.; Kirkwood, D. Small scale study of groundwater flow in a fractured carbonate-rock aquifer at the St-Eustache quarry, Québec, Canada. *Hydrogeol. J.* **2006**, *14*, 603–612.

17. Long, J.C.S.; Remer, J.S.; Wilson, C.R.; Witherspoon, P.A. Porous media equivalents for networks of discontinuous fractures. *Water Resour. Res.* **1982**, *18*, 645–658.

18. Hsieh, P.A.; Neuman, S.P.; Stiles, G.K.; Simpson, E.S. Field determination of the three-dimensional hydraulic

conductivity tensor of anisotropic media. 2 Methodology and application to fractured rocks. *Water Resour. Res.* **1985**,*21*, 1667–1676.

19. Neuman, S.P. Stochastic Continuum Representation of Fractured Rock Permeability as an Alternative to the Rev and Fracture Network Concepts. In Proceedings of the 28th US Symposium on Rock Mechanics, Tucson, AZ, USA, 29 June–1 July 1987; pp. 533–561.

20. Bradbury, K.R.; Muldoon, M.A.; Zaporec, A.; Levy, J. *Delineation of Wellhead Protection Areas in Fractured Rocks*; US EPA Technical Guidance Document, EPA 570/9-91-009; EPA: Washington, DC, USA, 1991.

21. Nuclear Regulatory Commission (NRC). *Rock Fractures and Fluid Flow: Contemporary Understanding and Applications*; National Academy Press: Washington, DC, USA, 1996.

22. Neuman, S.P. Trends, prospects and challenge in quantifying flow and transport through fractured aquifer. *Hydrogeol. J.* **2005**, *13*, 124–147.

23. Baiocchi, A.; Dragoni, W.; Lotti, F.; Piscopo, V. Sustainable yield of fractured rock aquifers: The case of crystalline rocks of Serre Massif (Calabria, Southern Italy). In *Fractured Rock Hydrogeology*; Sharp, J.M., Ed.; Taylor & Francis Group: London, UK, 2014; pp. 79–97.

24. Dragoni, W. Some considerations on climatic changes, water resources and water needs in the Italian region south of the 43° N. In *Water, Environment and Society in Times of Climatic Change*; Issar, A., Brown, N., Eds.; Kluwer Academic Publishers: Dordrecht, The Netherlands, 1998; pp. 241–271.

25. Intergovernmental Panel on Climate Change (IPCC). Summary for Policymakers. In *Climate Change 2013: The Physical Science Basis*; Cambridge University Press: Cambridge, UK, 2013.

26. Carmignani, L.; Cocozza, T.; Gandin, A.; Pertusati, P.C. Lineamenti della geologia dell'Iglesiente-Sulcis [Outlines of the geology of the Iglesiente-Sulcis Region]. In *Guida alla Geologia del Paleozoico Sardo*; Società Geologica Italiana, Guide Geologiche Regionali: Roma, Italy, 1982; pp. 65–77.

27. Istituto Superiore per la Protezione e la Ricerca Ambientale (ISPRA). *Notes to the Geological Map of Italy 1:50,000 "Sheet 566–Assemini"*; Servizio Geologico d'Italia: Roma, Italy, 2009. (In Italy)

28. Barca, S.; Cocozza, T.; del Rio, M.; Pittau Demelia, P. Discovery of lower Ordovician Acritarchs in the "Postgotlandiano" sequence of southwestern Sardinia (Italy): Age and tectonic implications. *Boll. Soc. Geol. Ital.* **1981**,*100*, 377–392.

29. Carmignani, L.; Conti, P.; Barca, S.; Cerbai, N.; Eltrudis, A.; Funedda, A.; Oggiano, G.; Patta, E.D. *Notes to the Geological Map of Italy 1:50,000 "Sheet 549–Muravera"*; Servizio Geologico d'Italia: Roma, Italy, 2001. (In Italy)

30. Celico, P.; Piscopo, V.; Berretta, G. Influence of a reservoir on groundwater flow in metamorphic rocks. *Geol. Appl. Idrogeol.* **1993**, *28*, 253–261. (In Italy).

31. Kruseman, G.P.; de Ridder, N.A. *Analysis and Evaluation of Pumping Test Data*; ILRI Publication 47; International Institute for Land Reclamation and Improvement: Wageningen, The Netherlands, 1994.

32. Allmendinger, R.W.; Cardozo, N.C.; Fisher, D. *Structural Geology Algorithms: Vectors & Tensors*; Cambridge University Press: Cambridge, UK, 2013.

33. Cardozo, N.; Allmendinger, R.W. Spherical projections with OSX Stereonet. *Comput. Geosci.* **2013**, *51*, 193–205.

34. Snow, D.T. Anisotropic permeability of fractured media. *Water Resour. Res.* **1969**, *5*, 1273–1289.

35. Oda, M. Permeability tensor for discontinuous rock masses. *Geotechnique* **1985**, *35*, 483–495.

36. Moye, D.G. Diamond drilling for foundation exploration. *Civil Eng. Trans. Inst. Eng.* **1967**, *9*, 95–100.

37. Theis, C.V. The relation between the lowering of the piezometric surface and the rate and duration of discharge of a well using groundwater storage. *Am. Geophys. Union Trans.* **1935**, *2*, 519–524.

38. Hantush, S.M.; Jacob, C.E. Non-steady radial flow in an infinite leaky aquifer. *Am. Geophys. Union Trans.* **1955**, *36*, 95–100.

39. Warren, J.E.; Root, P.J. The behavior of naturally fractured reservoirs. *One Petro* **1963**, *3*, 245–255.

40. Regione Sardegna (2010) Annali Idrologici. Available online: www.regione.sardegna.it (accessed on 25 October 2014).

41. Doherty, J. *PEST, Model-Independent Parameter Estimation–User Manual*, 5th ed.; Watermark Numerical Computing: Brisbane, Australia, 2010.

42. De Marsily, G.; Lavedan, C.; Boucher, M.; Fasaniso, G. Interpretation of interference tests in a well field using geostatistical techniques to fit the permeability distribution in a reservoir model. In *Geostatistics for Natural Resources Characterization*; NATO Advanced Study Institute: Dordrecht, The Netherlands, 1984; pp. 831–849.

43. Doherty, J. Groundwater model calibration using pilot-points and regularization. *Ground Water* **2003**, *41*, 170–177.

44. Clauser, C. Permeability of crystalline rocks. *EOS* **1992**, *73*, 233–240.

45. Schulze-Makuch, D.; Carlson, D.A.; Cherkauer, D.S.; Malik, P. Scale dependency of hydraulic conductivity in heterogeneous media. *Ground Water* **1999**, *37*, 904–919.

46. Renard, P.; de Marsily, G. Calculating equivalent permeability: A review. *Adv. Water Res.* **1997**, *20*, 253–278.

47. Neuman, S.P.; di Federico, V. Multifaceted nature of hydrogeologic scaling and its interpretation. *Rev. Geophys.* **2003**,*41*.

48. Krasny, J. Classification of transmissivity magnitude and variation. *Ground Water* **1993**, *31*, 230–236.

3

CHAPTER

Geological and geomorphological controls on groundwater occurrence in a hard rock region

S. P. Rajaveni, K. Brindha, L. Elango

Department of GeologyAnna University

ABSTRACT

Remote sensing and GIS tools have broadly helped hydrogeologists to delineate the groundwater prospective zones for watershed development and management. The origin, movement and existence of groundwater depends on several factors such as slope, drainage density, land use, geology, lineament density and geomorphology. Based on these, the mapping and identification of groundwater potential zones were carried out in a part of Nalgonda district, Telangana, India. The regions were categorised as high, moderate and low groundwater potential, and they were validated with the groundwater levels and yield of wells located in the corresponding zones. Extensive possibility for watershed development is possible in

41 % of the total 724 km^2 and 46 % of the area offers moderate options. Any groundwater management project implemented in these favourable areas will bring maximum benefit. Similar studies should be considered necessary before designing a water resource development activity as it will reduce the cost on detailed field visits which are time-consuming.

Keywords: Groundwater potential zones Overlay analysis Thematic layers Nalgonda Telangana India

INTRODUCTION

In a semi-arid country like India, surface water is not available throughout the year for various purposes. Due to meagre surface water resource, most of the requirements for irrigation, industry and domestic purposes are being met from groundwater. It is therefore essential to ensure the availability of groundwater throughout the year. Of many methods available, remote sensing is one of the techniques that can be used for rapid assessment of natural resources; however, it requires validation by field work. With the advances and availability of satellite images, it is possible to indirectly identify the ground conditions through the surface and subsurface features such as topography, land use, drainage, geology and geomorphology.

Remote sensing and geographic information system (GIS) have proved to be important tools to prepare the thematic maps from satellite images and is also used as a multi-criteria decision analysis tool (Vittala et al. 2005; Madrucci et al. 2008; Mondal et al. 2008; Javed and Wani2009; Jha et al. 2010; Dar et al. 2010). Application of remote sensing and GIS in geology and geomorphology for quick hydrogeological evaluation has been proved to be efficient (Pareta and Pareta 2011; Suganthi et al. 2013; Anbazhagan and Jothibasu 2014; Ndatuwong and Yadav2014). Interpretation of various features has also been done by visual method, which is time-consuming and difficult in areas where accessibility is not possible to humans. To overcome these constraints, integrating remote sensing and GIS methods gives more accuracy and reduces the risk of manual error (Pinder 2002; Prasad et al. 2008; Dar et al. 2011; Jasmin and Mallikarjuna 2011). Another advantage is that both discrete and continuous data from primary and secondary sources can be combined and interpreted by this method (Burrough 1986). Rai et al. (2005), Karunanidhi et al. (2013), Rashid et al. (2012) and Srinivasan et al. (2013) complemented remote sensing methods with geoelectrical studies to identify the groundwater prospective zones. Researchers have also used statistical methods (Ozdemir 2011; Machiwal et al. 2011; Manap et al. 2012, 2013; Pinto et al. 2015) and numerical modelling (Chenini and Mammou 2010; Gaur et al. 2011; Ahmad et al. 2011) along with GIS for groundwater potential mapping. Ideally, GIS is a rapid and cost-effective problem-solving platform for identifying groundwater potential zones. If this method is complemented with necessary ground checks, it will be an effective technique in hydrogeological studies. This has been agreed broadly in groundwater assessment studies, and several researchers have identified the groundwater potential zones in various regions by the integration of remote sensing and GIS (Sener et al.2005; Varughese et al. 2012; Lone et al. 2013; Narendra et al. 2013).

Nalgonda district in Telangana, southern India, is intensively irrigated, and groundwater (57.2 %) is the major source of supply for agriculture (CGWB 2007). The groundwater quality of this area has been studied earlier (Brindha and Elango 2013a, b; Brindha and Elango 2012; Elango et al. 2012; Rajesh et al. 2012; Brindha et al. 2011; Brindha and Elango 2010). In spite of the large-scale use of groundwater in this area, it has not been classified based on groundwater potential and a study of this nature is essential to sustainably utilise the available groundwater resources and to meet the growing needs for agricultural, domestic and industrial activities. Hence, this study was carried out with an objective to map the groundwater potential zones in a part of Nalgonda district with the help of thematic maps prepared using remote sensing and GIS. The results of this study were validated using the groundwater level monitored in these areas.

DATA COLLECTION

Toposheet was used as a reference for preparing the thematic maps. Survey of India toposheets (1:25,000) covering the study area was used to prepare the base map, drainage and topographical maps. Geological map of Geological Survey of India (GSI), Hyderabad, was used to identify the different geological features of this area (GSI 1995). This was further updated by ground-truth verification carried out by geological fieldwork along different traverses across the area in July 2010. Vertical electrical sounding (VES) was carried out in 37 locations of the study area to understand the geology and also to verify the extent of weathering. Geomorphology and land use of the area were delineated from the Indian Remote Sensing Satellite

P6 (Resourcesat-1) Linear Imaging Self-scanning Sensor IV (IRS P6 LISS IV) images of the year 2008 with a spatial resolution of 5.8 m by visual interpretation technique. Figure 1 shows the steps adopted for this study. For the verification process, groundwater levels were measured in forty-two monitoring wells in this area once in 2 months from March 2008 to January 2009 using a water level indicator (Solinst 101). Global positioning system (GPS) was used to measure the topographical elevation at the groundwater level measuring location. Further five pumping tests were also carried out to determine the yield of wells to validate the groundwater potential zones.

Description of the study area

The study area (Fig. 2) (724 km^2) is located in Nalgonda district, Telangana, India. This area is situated at a distance of about 80 km ESE of Hyderabad, the capital of Telangana state. Nagarjuna Sagar reservoir is present at the south-eastern side of the study area. The southern boundary is bounded by Pedda Vagu River and the northern boundary is partly bounded by Gudipalli Vagu River. The climate in this area is arid to semi-arid. The temperature ranges from 30 to 46.5 °C during summer (April–June) and between 17 and 38 °C during winter (November–January). South-west monsoon occurs during June–September resulting in an average annual rainfall of about 600 mm.

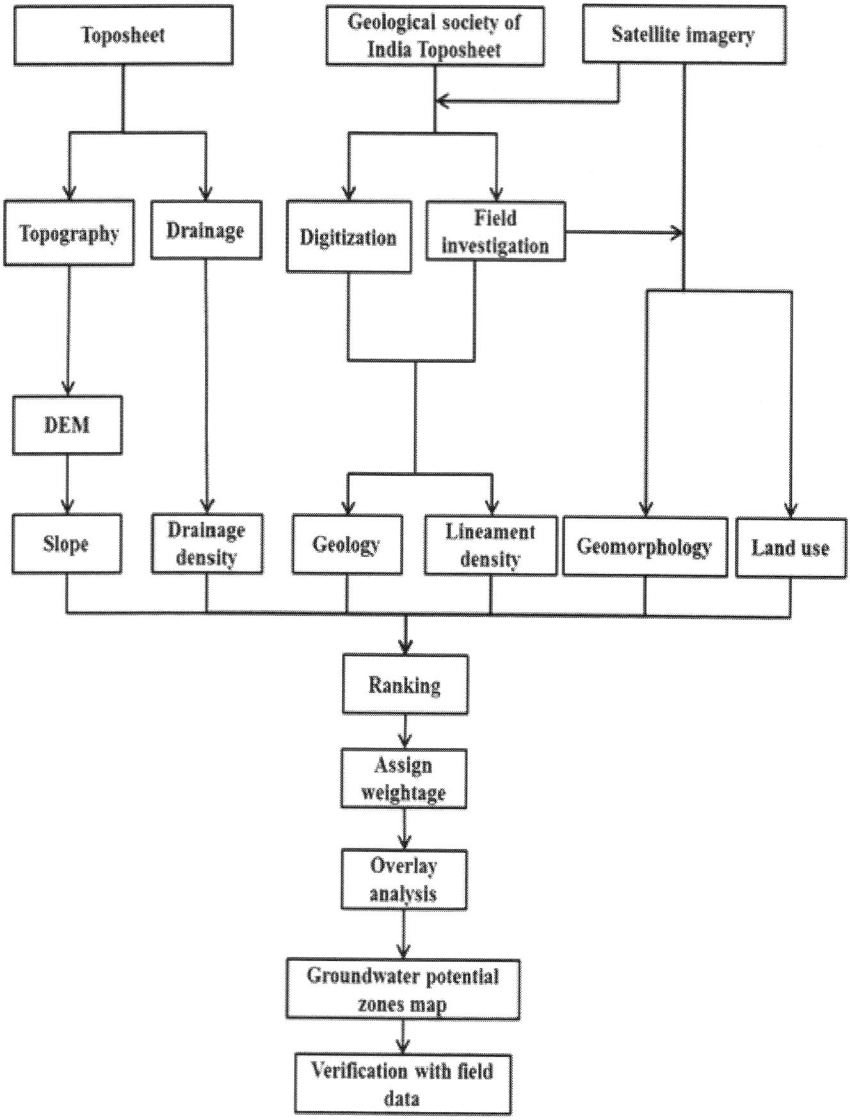

Fig. 1 Methodology adopted for identification of groundwater potential zones

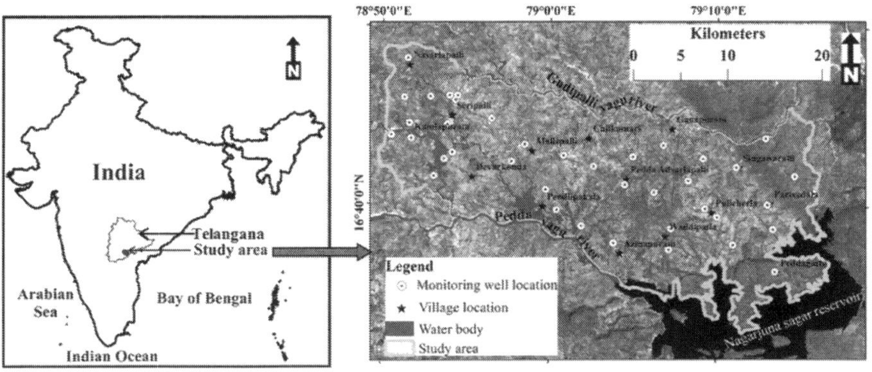

Fig. 2 Location of the study area

RESULTS AND DISCUSSIONS

The quality of a study of this nature will improve depending on the number of accurate thematic layers that have been used. Depending on the availability of data and resources, this study is limited to and concentrates on the use of slope, drainage density, land use, geology, lineament density and geomorphological characteristics. Table 1 gives the area covered by various features of a spatial layer, and Table 2 gives the ranks and weights assigned to the features based on literature.

Slope

The influence of slope will have a remarkable effect on the infiltration of surface water from the ground. A low slope percentage indicates that the surface water received mainly because of precipitation will have more time to remain on the ground surface and percolate into the subsurface. In case of highly sloping areas, the run-off is more immediate offering less retention time for the water on the ground surface and significantly reduces the groundwater recharge potential.

As the slope greatly depends on the elevation in an area, to determine this characteristic, the topography should be known.

Table 1: Area covered under different range of parameters

Parameter	Range	Area in km^2	Percentage
Slope	1°–5°	674.5	93.2
	5°–20°	33.9	4.7
	20°–80°	15.6	2.2
Land use	Water body	57.5	7.9
	Waste land	140.3	19.4
	Forest	56.5	7.8
	Built-up	18.5	2.6
	Agriculture	451.2	62.3
Geomorphology	Denudational hill	36.6	5.1
	Linear ridge	7.7	1.1
	Pediments	39.5	5.5
	Pediplain	560.4	77.4
	Water body	11.1	1.5
	Residual hill	10.1	1.4
	Structural hill	48.1	6.6
	Valley fill	10.7	1.5
Geology	Quartzite	22.8	3.2
	Quartz vein	1.5	0.2
	Pink biotite granite	36.4	5.0
	Migmatite granite	75.4	10.4
	Meta-basalt	10.2	1.4
	Meta-rhyolite	0.6	0.1
	Grey biotite granite	557.4	77.0
	Grey hornblende biotite	10.5	1.5
	Grey biotite gneiss	7.2	1.0
	Dolerite	2.0	0.3
Drainage density (km^{-1})	0–0.75	243.4	33.6
	0.75–2.39	345.8	47.8
	2.39–5.08	134.9	18.6
Lineament density (km^{-1})	0–0.37	468.7	64.7
	0.37–1.08	175.6	24.3
	1.08–2.96	79.7	11.0

Table 2: Ranks and weightages assigned to various thematic maps for delineation of groundwater potential zones

Thematic layer	Sub-layers/class	Rank (high = 4; low = 1)	Weightage of each layer	References
Slope	Low slope (0°–5°)	3	0.1	Jasmin and Mallikarjuna (2011), Mondal et al. (2008), Bagyaraj et al. (2013), Lone et al. (2013) and Rashid et al. (2012)
	Medium slope (5°–20°)	2		
	High slope (>20°)	1		
Land use	Waterbody	3	0.15	Kumar et al. (2008), Lone et al. (2013), Nagaraju et al. (2011), Pandian and Kumanan (2013), Varughese et al. (2012), Sener et al. (2005)
	Agriculture	2		
	Wasteland, built-up, forest	1		
Geomorphology	Water body (reservoir)	3	0.25	Dar et al. (2010), Kumar et al. (2008), Bagyaraj et al. (2013), Lone et al. (2013), Pandian and Kumanan (2013)
	Valley fill, pediplain	2		
	Pediments, denudational hill, residual hill, linear ridge and structural hill	1		
Geology	Meta-basalt	3	0.1	Jasmin and Mallikarjuna (2011), Kumar et al. (2008), Lone et al. (2013)
	Quartz vein, grey hornblende biotite and grey biotite gneiss	2		
	Meta-rhyolite, dolerite dyke, grey biotite granite, migmatite granite, pink biotite granite and quartzite	1		
Drainage density	Low density (0–0.75 km^{-1})	3	0.2	Jasmin and Mallikarjuna (2011), Mondal et al. (2008), Bagyaraj et al. (2013), Rashid et al. (2012)
	Moderate density (0.75–2.39 km^{-1})	2		
	High density (2.39–5.08 km^{-1})	1		
Lineament density	Low density (0–0.37 km^{-1})	1	0.2	Jasmin and Mallikarjuna (2011), Dar et al. (2010), Mondal et al. (2008), Rashid et al. (2012)
	Moderate density (0.37–1.08 km^{-1})	2		
	High density (1.08–2.96 km^{-1})	3		

Topography of the study area delineated from the toposheets (1:25,000 scale) comprises of three distinct features. The Nagarjuna Sagar reservoir in the south-eastern hemicircle, the Lambapur–Peddagattu and Seripalli plateaus and the gently sloping undulating land make up the varied topographical characteristics of this area. The extreme south-east occupied by the Nagarjuna Sagar reservoir has maximum water level at 180 mamsl. There are several small hills in this area with height ranging from 250 to 300 mamsl. The Lambapur and Peddagattu plateaus occur as flat topped hills in the south-east of the study area. The highest topographical elevation of this area is 348.1 mamsl and lowest is 169.5 mamsl. The ground surface generally slopes towards the south-east direction. DEM was generated from 10-m interval contour of the topographical map, which is given in Fig. 3a. Slope varied from 1° to 80° (Fig. 3b), and most parts of the area have slope ranging from 1° to 2.5° (Table 1). The south-east and north-west regions of the study area have a slope ranging from 2.5° to 80°. For the assessment of groundwater potential zones, this area was classified based on three ranges of slope (Fig. 3b). The range with lesser slope (0°–5°) was given a higher rank of 4 as it increases the infiltration rate (Lone et al. 2013), whereas a higher slope from 20° to 80° was given a low rank of 1 (Bagyaraj et al. 2013).

Land use

The dependence on groundwater especially quantitatively is determined by the land use in an area. Figure 4 shows that the land use in this area which comprises of agricultural land (cropland, fallow land and plantation), water bodies (stream and tank), forest plantation, wasteland (land with or without scrub, stony waste, rocky knob and sheet rock area) and built-up area as identified from LISS IV imagery. Area occupied by different land-use characteristics is given in Table 1. Irrigation is mainly (57 %) dependent on groundwater (CGWB 2007) apart from rainfall. The cropping pattern depends on the climatic conditions and availability of water sources. Paddy, sweet lime, castor, cotton, grams and groundnut are the major crops grown in this area. Majority of the livelihood is from agriculture and its allied activities.

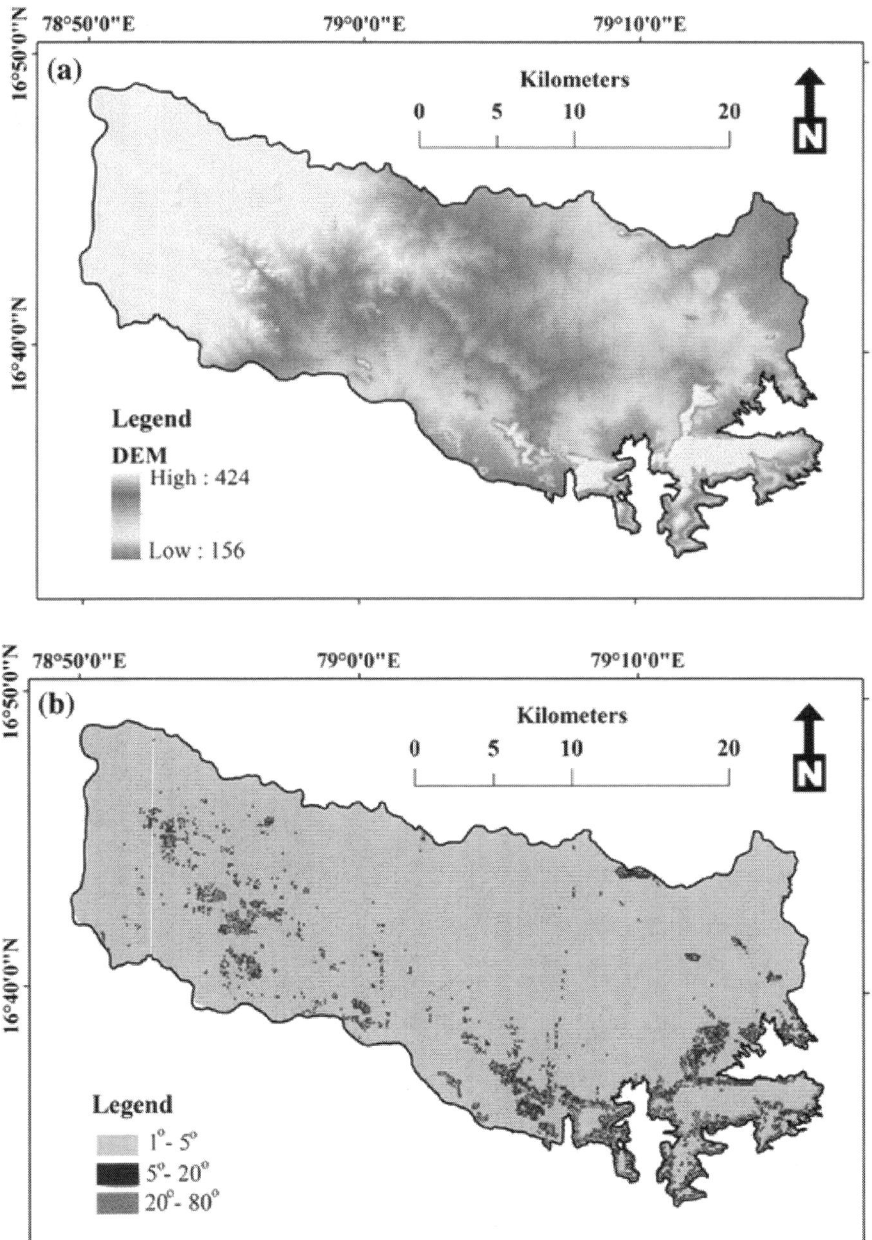

Fig. 3 a Topography (mamsl), **b** slope characteristics

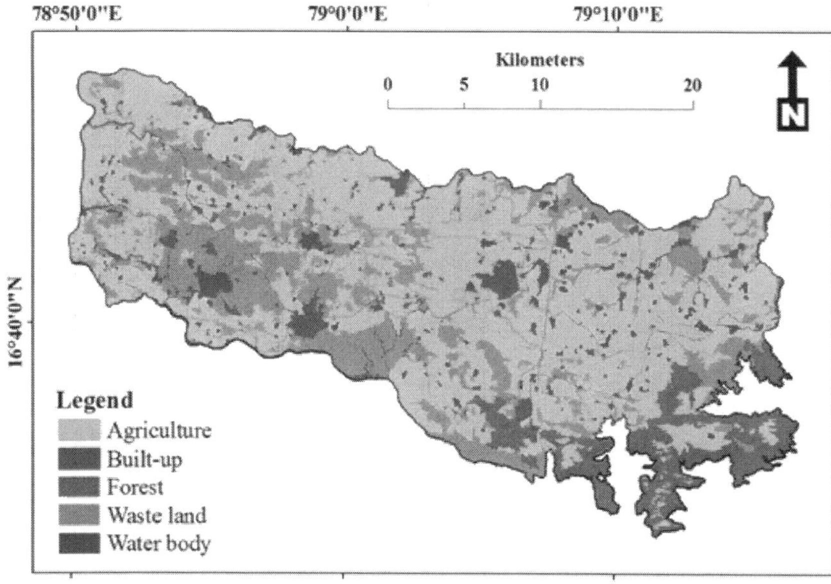

Fig. 4 Land use

The demarcation of different land use in the study area was derived from false colour composite (FCC) IRS LISS IV images by visual interpretation techniques (Fig. 4). The crop land was identified by light medium red tone and fine/medium texture; fallow land was identified by medium burly wood tone and medium texture; plantation was identified by dark red tone and fine texture, respectively (Lone et al. 2013). These are classified as good potential sites for groundwater prospects (Varughese et al. 2012). Water bodies were displayed in black tone and were assigned as high potential zones with a rank of 3 (Chipman and Lillesand 2007). Forest plantations had light reddish brown tone and fine medium texture with irregular shape and varying size (Kumar et al. 2008). Groundwater potential in forest areas was assigned as poor because even though these areas may have good groundwater recharge, the groundwater is not being extracted from this land (Magesh et al. 2012). Wasteland was identified by light to dark bluish tone with coarse texture, and built-up land was displayed as light bluish white with fine texture of regular shape and size (Lillesand et al. 2007). These areas were also assigned low values of 1 for mapping the groundwater potential zones (Table 2).

Geology

Geological mapping serves as a basis for the study of groundwater condition in an area. Particular attention is to be paid to the geological units that affect the occurrence, movement and quality of groundwater. Initially, geological map of the region was digitised from GSI map (GSI 1995), which was updated by ground-truth verification carried out in the study area in March 2008. Further the IRS LISS IV data were also used to improve the geological map. Subsequently, another field work was carried out in July 2010 to verify and validate the prepared geological map.

The basement of this area is constituted by medium- to coarse-grained granite (Fig. 5), which comprises the Srisailam formation, the youngest member of Cuddapah super group. Grey and pink granites are exposed in most parts of the area. These rocks have less porosity and are classified into low groundwater potential zones with low rank. Migmatite granite occurs at north-eastern side and the south-eastern part is underlain by Srisailam metamorphic formations. Arenaceous and pebbly–gritty quartzite shale with dolomite limestone, intercalated sequence of shale quartzite and massive quartzite sediments are also present in the Srisailam formations. The stratigraphic sequence occurring in the study area is given in Table 3, and the area occupied by each geologic feature is given in Table 1. Generally hard rock does not have primary porosity. Secondary porosity in the form of fractures and weathering only allow groundwater occurrence (Vittala et al. 2005). The extent of weathering was identified based on VES which indicated weathering up to a depth of 40 m from the ground surface. High rank is assigned for meta-basalt rock (Table 2), which has high amount of porosity and good groundwater potential.

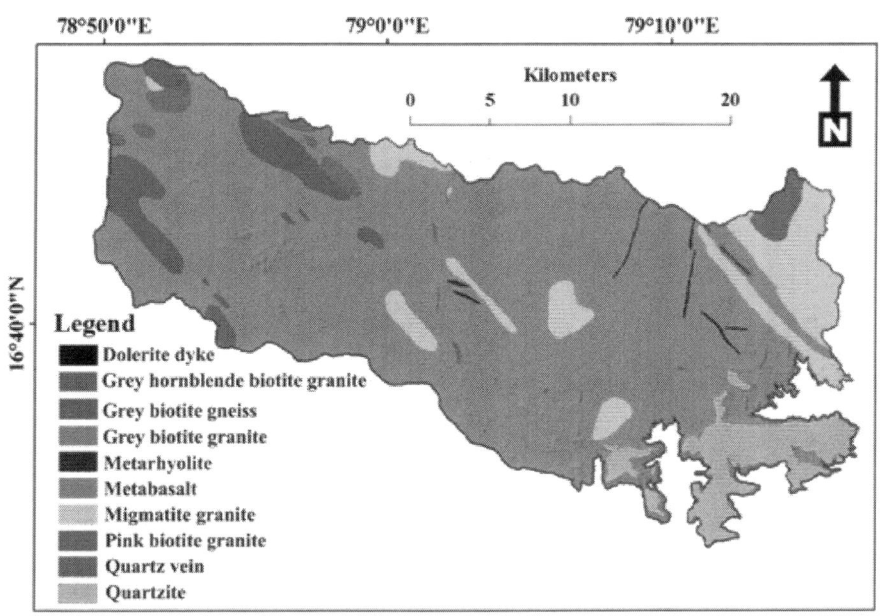

Fig. 5 Geology

Table 3 Stratigraphic sequence of the study area (after GSI 1995)

Cuddapah Super Group		Massive quartzite
	Srisailam Formation	Upper shale
		Quartzite with shale intercalation
		Lower shale with limestone intercalation
		Pebbly and gritty quartzite/ arenite
		Uranium mineralized region
--------------------------------	Unconformity	--------------------------------
Late Archean/ Lower Proterozoic		Granite/ granitic gneiss with intrusion of dolerite dykes and quartz veins

Geomorphology

The geometric features of different landforms are created by the dynamic action of several geological processes such as temperature changes, freezing and thawing, chemical reactions, seismic shaking and moving of wind and water. Denudational hill, linear ridge, pediments, pediplain, water body, residual hill, structural hill and valley fill were identified from IRS P6 LISS IV image through visual interpretation (Fig. 6). The area occupied by different geomorphological features in the study area is given in Table 1.

Erosion and weathering process creates denudational hill and is poor to infiltration of surface water (Ramaiah et al. 2012). Linear ridges are identified as dark brown tone linear structures. Pediment is the gentle sloping surface formed between hill and plain surface. These appear as coarse tone and texture in FCC satellite image and have poor groundwater potential. Pediplain appears in light to dark red

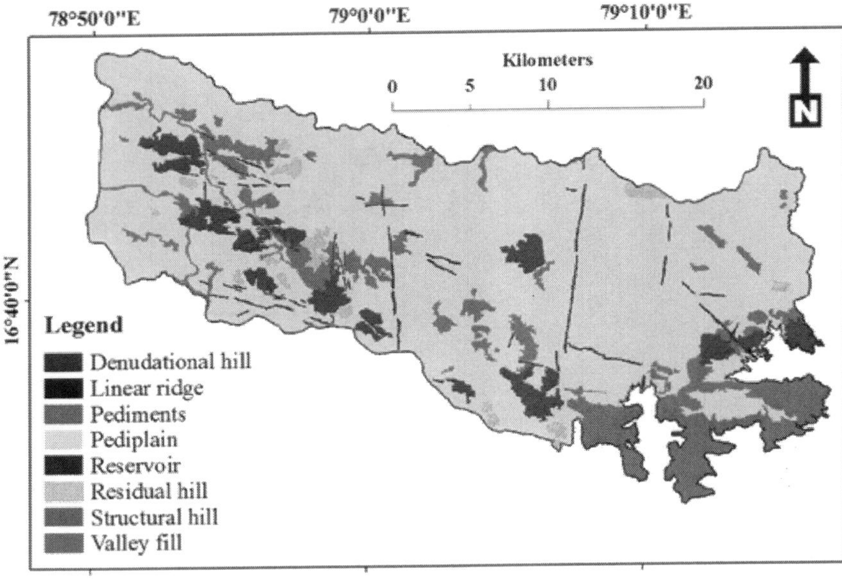

Fig. 6 Geomorphology

tone, fine to coarse texture, wide spread size and irregular shape which are associated with agricultural land that yield moderate to good groundwater. In satellite images, residual hill displays brown to dark red tone, coarse texture and large size while structural hill which also appears as brown to dark red tone has coarse texture, small size and are irregular in shape. Valley fill is developed due to the deposition of transported and weathered materials in valley areas that are generally controlled by lineaments. It exhibits dark red to bright red tone on FCC image. Residual hill, structural hill and linear ridge landforms are considered as very low groundwater potential as they are unfractured rock and have low infiltration (Ramaiah et al. 2012). Pediplain, valley fill and water bodies allow high infiltration and have good groundwater potential, hence, are assigned a higher rank (Table 2).

Drainage density

Drainage in an area depends on the topography, slope and the subsurface characteristics. Drainage density is calculated as the total length of all the streams and rivers in a basin divided by the total area of the drainage basin. The higher the drainage density, the higher the run-off will be, and infiltration of water into the subsurface is affected.

This area is bounded by two major rivers, Pedda Vagu in the south and partly by Gudipalli Vagu in the north. These two rivers are seasonal and flow during the south-west monsoon falling mainly between July and September. Drainage pattern is mainly dendritic to sub-dendritic (Fig. 7a), whereas trellis-type drainage pattern is also seen in some parts of the area. Many tanks and few small reservoirs are present in the depressed parts of the undulating topography. This area also has wide-lined canal networks catering for irrigation purposes. Drainage density calculated by Kernel density method to understand the potential of the watershed to favour the groundwater potential is given in Fig. 7b. Most of the area falls under 0.75 to 2.39 km^{-1}drainage density (Table 1). The drainage density map was reclassified with areas having less density (0–0.75 km^{-1}) designated with higher rank (Table 2) as they facilitate recharge (Bagyaraj et al. 2013), and areas with high density (2.39–5.08 km^{-1}) were designated with lower rank (Table 2) as they facilitate less infiltration and recharge (Rashid et al. 2012).

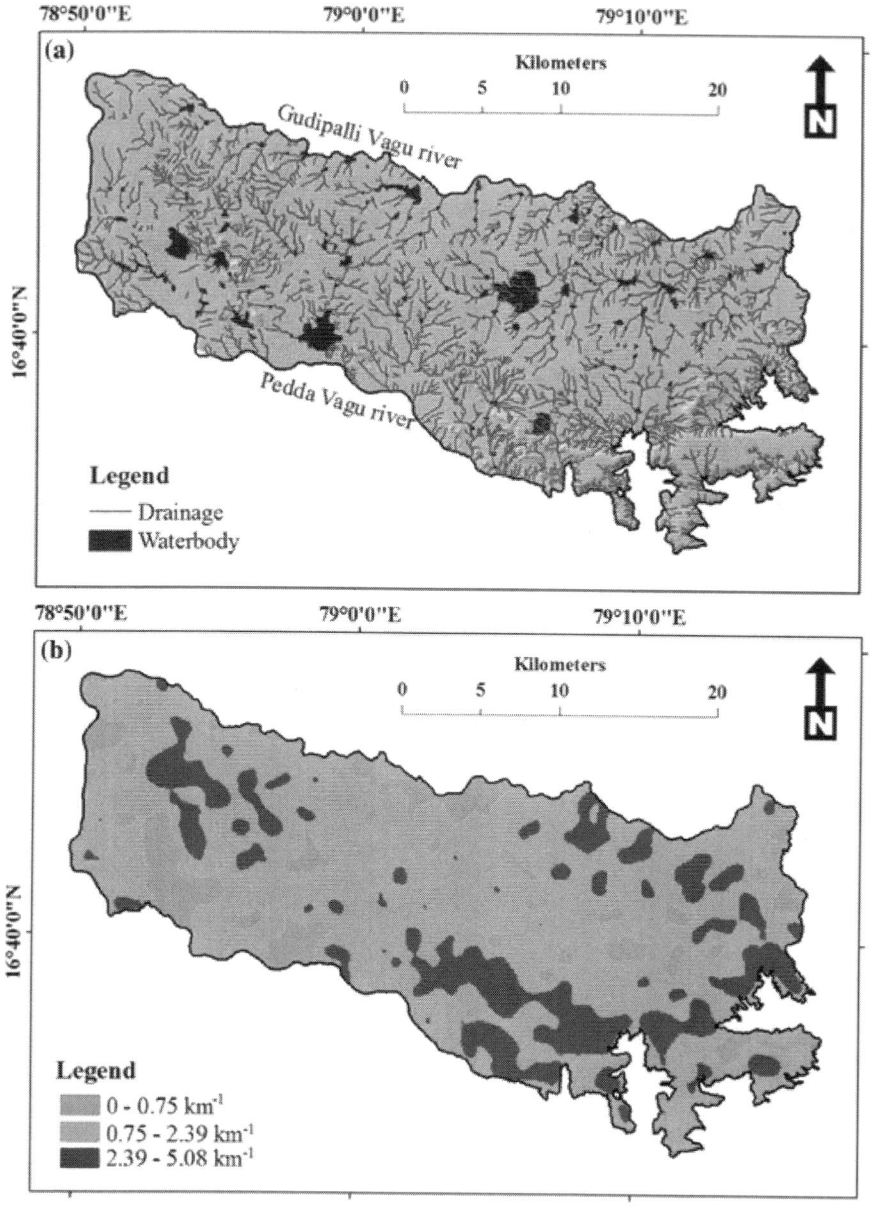

Fig. 7 a Drainage and **b** drainage density

Lineament density

Developed by the tectonic activity, lineaments reflect a general surface manifestation of underground fractures, with inherent characteristics of porosity and permeability of the underlying materials (Rao 2006). These features serve as the main conduits for movement and storage of groundwater in impermeable rocks worldwide (Preeja et al. 2011) in the form of dykes and faults. They were delineated from geological map, satellite imagery and field visits. Fractures seen on the outcrops, excavations and well sections were studied during the field visits, and most of the fractures are trending NNE–SSW and NS (Rajaveni et al. 2014). There are 217 intrusive dolerite dykes and 5 faults which have been identified and mapped. In general, the length of these dykes ranges from 0.17 to 10.23 km, whereas most of the dykes are of length ranging between 0.1 and 1.5 km. From field investigation, it is observed that all the dykes are vertically dipping.

Lineament density map was prepared by using line density method (Fig. 8). The total area falling in different range of lineament density is given in Table 1 which shows that most parts of the area have low lineament density. Around 90 % of lineaments are dykes in this area which might affect groundwater flow and have low groundwater potential. But these dykes are highly weathered and do not act as barrier for groundwater flow which has also been reported by Rajesh et al. (2012) and Rajaveni et al. (2014). So, low lineament density indicates low infiltration rate, whereas the high lineament density areas indicate high infiltration (Table 2) thus a potential zone for groundwater development.

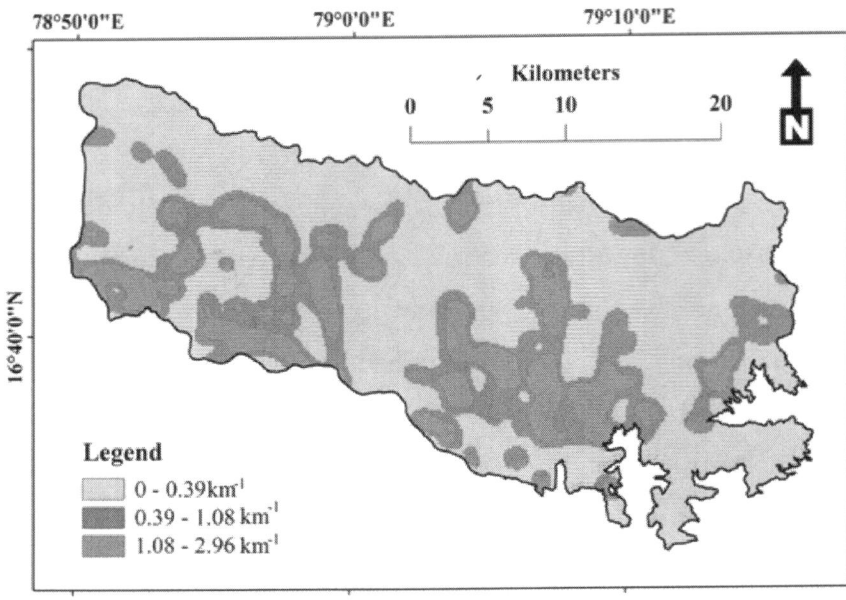

Fig. 8 Lineament density

Groundwater potential zonation and validation

Though several features are used to identify the groundwater prospects, this study limits itself to five features to evaluate the groundwater prospects through satellite-based remote sensing and GIS in a hard rock terrain. The various distinct characteristics and sub-features in each of these thematic layers were reclassified by assigning ranks for computational simplicity. Later, each of these thematic layers were assigned suitable weights based on several studies which is given in Table 2. They were then integrated to bring out a groundwater potential index value based on the following formula.

$$\text{Groundwater potential index} = (S_w \times S_r) + (\text{DD}_w \times \text{DD}_r) + (\text{LU}_w \times \text{LU}_r) + (G_w \times G_r) + (\text{LD}_w \times \text{LD}_r) + (\text{GM}_w \times \text{GM}_r)$$

where S = slope; DD = drainage density; LU = land use; G = geology; LD = lineament density; GM = geomorphology; w = feature weight; r = ranking.

The final map (Fig. 9) prepared using the analysis tool brought out a range of value from 1 to 2.8. This was further classified into three categories as low (1–1.45), moderate (1.5–1.85) and high (1.9–2.8). About 41.3 % (299 km^2) of the area is classified as high groundwater potential, while some parts in the south-east of the study area have low groundwater prospects (12.5 %). Moderately suitable area for groundwater development was distributed throughout this region occupying 334 km^2 of the area (46.1 %).

High groundwater potential areas are present in the agricultural areas with low slope and less drainage density. Lineaments in this area do not affect groundwater potential and high groundwater potential zones occur surrounding high lineament density areas. Groundwater in this area is restricted to geomorphological features of pediplains, valley and pediments. Regions with low groundwater potential area are spread mainly in regions with quartzite and dolerite dyke and with hydrogeomorphological units such as denudational hill, linear ridge, residual hill and structural hill. In general, low slope, low drainage density, low lineament density associated with gneiss are highly promising area for groundwater development. The complex geology, geomorphology and land use result in varied groundwater potential zones of this area. A general observance of the prospective zones based on the various features is given in Table 4.

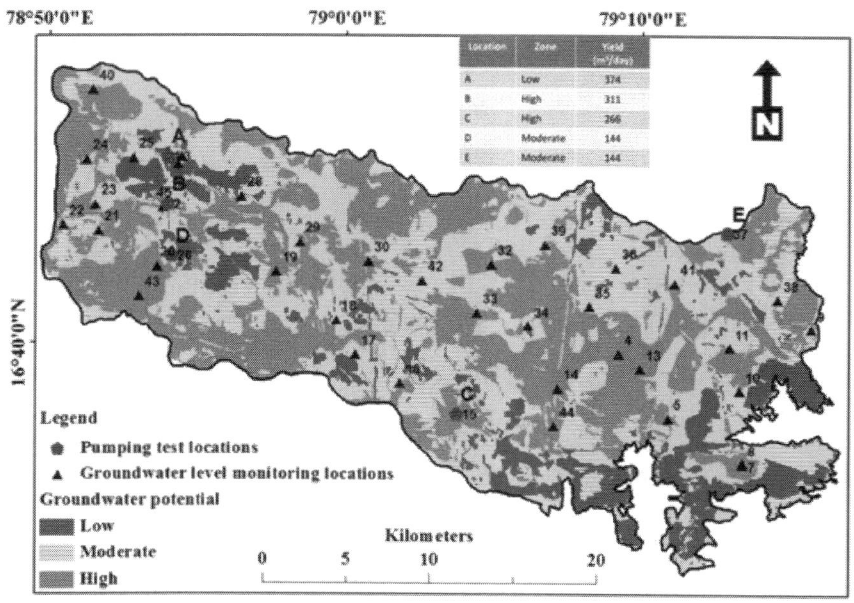

Fig. 9 Groundwater potential zones in a part of Nalgonda district, Telangana, with yield of wells

Groundwater potential zonation map prepared was validated with the groundwater levels measured. Groundwater levels in 42 wells which were more or less uniformly distributed over this area were measured once in 2 months as a part of this study between March 2008 and January 2009. Measurement of groundwater level fluctuation was used to determine the range of variation during the

Table 4 Features of high, moderate and low groundwater potential zones

Feature	High	Moderate	Low
Slope	0°–5°	0°–20°	5°–20°/>20°
Land use	Agriculture, water body	Built-up, wasteland, forest	Built-up, wasteland, forest
Geomorphology	Pediplain, water body, valley	Pediments	Denudational hill, linear ridge, residual hill, structural hill
Geology	Grey biotite gneiss, grey biotite granite	Grey biotite granite, meta-rhyolite, meta-basalt, migmatite granite	Migmatite granite, dolerite dyke, quartzite
Drainage density	0–2.39 km^{-1}	0–2.39 km^{-1}	2.39–5.08 km^{-1}
Lineament density	0–0.39/0.39–1.08 km^{-1}	0.39–1.08/1.08–2.96 km^{-1}	1.08–2.96 km^{-1}

study period. It is observed that the groundwater level fluctuation is comparatively less in the monitoring wells located in zones that are classified as high groundwater potential (Fig. 10). In contrast, in the wells that are located in moderate groundwater potential zones the fluctuation in groundwater level was high. In three wells located in the low potential zones, moderate groundwater level fluctuation was observed. However, there are a few exceptions in certain cases which may be due to localised variation in hydrogeological conditions and pattern of groundwater pumping from these wells. Yield of the wells assessed based on pumping tests is shown in Fig. 9 along with groundwater potential zones. Three of the wells were located in high potential zone and two in moderate potential zone, and the yield was higher in the former compared to the latter (Fig. 9). However, the pumping test carried out in well 'A' located in low groundwater potential zone indicated a high well yield of value 374 m^3/day. This may be due to the possible presence of well intersecting connected fractures. Thus, except for this location there was reasonable comparison between the well yield and the zones of groundwater potential. These analyses confirm that groundwater potential zone map derived by remote sensing and GIS method is reliable for this region. The findings from this study prove to be useful and can be used for groundwater development and governance activities in this area. However, as indicated by a few exceptions in comparison of groundwater level fluctuation and well yield with potential zones, greater care need to be used on the application of remote sensing techniques since it is difficult to understand the fracture geometry on a regional scale and consider the same in GIS techniques.

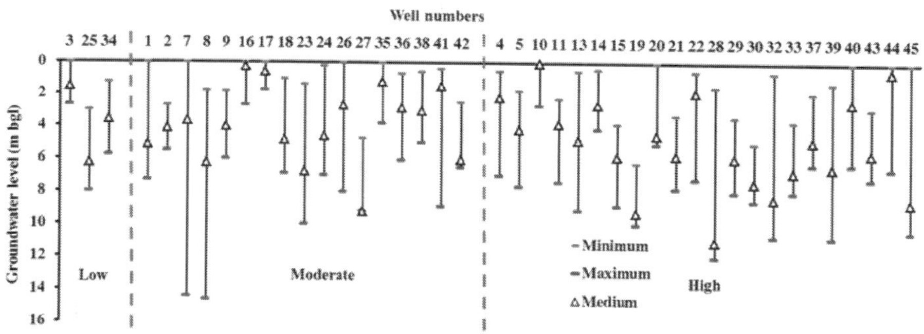

Fig. 10 Groundwater level fluctuations in different groundwater potential zones

CONCLUSION

Mapping and identifying the groundwater potential zones in an area help to plan for sustainable groundwater exploitation and management. As agriculture is the main activity in this part of Nalgonda district, India, the demand for groundwater resource is high. Conducive areas for groundwater occurrence and storage were identified based on slope, drainage density, land use, geology, lineament density and geomorphological features of this area. The best suitable areas identified were mainly in pediplain and valleys with minimum slope. Areas with agricultural practice provided wide possibility for groundwater development with high lineament density and low drainage density. Most of the area (299.3 km^2) posed good groundwater recharge capacity. The areas identified based on integrated mapping of various features with intensive geological and geomorphological mapping were verified by field observation of groundwater level data which proved to be comparable. Favourable groundwater potential zones identified from this study will serve as a tool to various organisations and decision makers in planning for groundwater resource exploration and development in this area which will result in high yield.

ACKNOWLEDGMENTS

The authors acknowledge the Board of Research in Nuclear Sciences, Department of Atomic Energy, Government of India for funding this work (Grant No. 2007/36/35). Authors also thank the Department of Science and Technology's Funds for Improvement in Science and Technology scheme (Grant No. SR/FST/ESI-106/2010) and University Grants Commission's Special Assistance Programme [Grant No. UGC DRS II F.550/10/DRS/2007(SAP-1)] for their support in creating laboratory facilities which helped in carrying out part of this work.

REFERENCES

1. Ahmad Z, Ashraf A, Fryar A, Akhter G (2011) Composite use of numerical groundwater flow modeling and geoinformatics

techniques for monitoring Indus Basin aquifer, Pakistan. Environ Monit Assess 173(1–4):447–457

2. Anbazhagan S, Jothibasu A (2014) Geoinformatics in groundwater potential mapping and sustainable development: a case study from southern India. Hydrol Sci J.

3. Bagyaraj M, Ramkumar T, Venkatramanan S, Gurugnanam B (2013) Application of remote sensing and GIS analysis for identifying groundwater potential zone in parts of Kodaikanal Taluk, South India. Front Earth Sci 7(1):65–75

4. Brindha K, Elango L (2010) Study on bromide in groundwater in parts of Nalgonda district, Andhra Pradesh. Earth Sci India 3(1):73–80

5. Brindha K, Elango L (2012) Groundwater quality zonation in a shallow weathered rock aquifer using GIS. Geo Spat Inf Sci 15(2):95–104

6. Brindha K, Elango L (2013a) Causes for variation in bromide concentration in groundwater of a granitic aquifer. Int J Res Chem Environ 3(2):163–171

7. Brindha K, Elango L (2013b) Occurrence of uranium in groundwater of a shallow granitic aquifer and its suitability for domestic use in southern India. J Radioanal Nucl Chem 295(1):357–367

8. Brindha K, Elango L, Nair RN (2011) Spatial and temporal variation of uranium in a shallow weathered rock aquifer in southern India. J Earth Syst Sci 120(5):911–920

9. Burrough PA (1986) Principles of geographical information systems for land resources assessment. Oxford University Press, Oxford, p 193

10. CGWB (Central Ground Water Board) (2007) Ground water information Nalgonda district Andhra Pradesh, India

11. Chenini I, Mammou AB (2010) Groundwater recharge study in arid region: an approach using GIS techniques and numerical modeling. Comput Geosci 36(6):801–817

12. Chipman JW, Lillesand TM (2007) Satellite-based assessment of the dynamics of new lakes in southern Egypt. Int J Remote Sens 28(19):4365–4379

13. Dar MA, Sankar K, Dar IA (2010) Groundwater prospects evaluation-based on hydrogeomorphological mapping: a case study in Kancheepuram district, Tamil Nadu. J Indian Soc Remote Sens 38(2):333–343

14. Dar IA, Sankar K, Dar MA (2011) Deciphering groundwater potential zones in hard rock terrain using geospatial technology. Environ Monit Assess 173(1–4):597–610

15. Elango L, Brindha K, Kalpana L, Sunny F, Nair RN, Murugan R (2012) Groundwater flow and radionuclide decay-chain

transport modelling around a proposed uranium tailings pond in India. Hydrogeol J 20(4):797–812

16. Gaur S, Chahar BR, Graillot D (2011) Combined use of groundwater modeling and potential zone analysis for management of groundwater. Int J Appl Earth Obs Geoinf 13(1):127–139

17. GSI (Geological Survey of India) (1995) Geology and minerals map of Nalgonda district. Andhra Pradesh, India

18. Jasmin I, Mallikarjuna P (2011) Review: satellite-based remote sensing and geographic information systems and their application in the assessment of groundwater potential, with particular reference to India. Hydrogeol J 19(4):729–740

19. Javed A, Wani MH (2009) Delineation of groundwater potential zones in Kakund watershed, eastern Rajasthan, using remote sensing and GIS techniques. J Geol Soc India 73(2):229–236

20. Jha MK, Chowdary VM, Chowdhury A (2010) Groundwater assessment in Salboni Block, West Bengal (India) using remote sensing, geographical information system and multi-criteria decision analysis techniques. Hydrogeol J 18(7):1713–1728

21. Karunanidhi D, Vennila G, Suresh M, Karthikeyan P (2013) Geoelectrical Schlumberger investigation for characterizing the hydrogeological conditions using GIS in Omalur Taluk, Salem District, Tamil Nadu, India. Arab J Geosci 7:1791–1798

22. Kumar MG, Agarwal AK, Bali R (2008) Delineation of potential sites for water harvesting structures using remote sensing and GIS. J Indian Soc Remote Sens 36(4):323–334

23. Lillesand T, Kiefer RW, Chipman J (2007) Remote sensing and image interpretation. Wiley, Hoboken

24. Lone MS, Nagaraju D, Mahadavesamy G, Siddalingamurthy S (2013) Applications of GIS and remote sensing to delineate artificial recharge zones (DARZ) of groundwater in H.D. Kote taluk, Mysore district, Karnataka, India. Int J Remote Sens Geosci 2(3):92–97

25. Machiwal D, JhaM K, Mal BC (2011) Assessment of groundwater potential in a semi-arid region of India using remote sensing, GIS and MCDM techniques. Water Resour Manage 25(5):1359–1386

26. Madrucci V, Taioli F, deAraújo CC (2008) Groundwater favorability map using GIS multicriteria data analysis on crystalline terrain, São Paulo State, Brazil. J Hydrol 357(3–4):153–173

27. Magesh NS, Chandrasekar N, Soundranayagam JP (2012) Delineation of groundwater potential zones in Theni district, Tamil Nadu, using remote sensing, GIS and MIF techniques. Geosci Front 3(2):189–196

28. Manap MA, Nampak H, Pradhan B, Lee S, Sulaiman WNA, Ramli MF (2012) Application of probabilistic-based frequency ratio

model in groundwater potential mapping using remote sensing data and GIS. Arab J Geosci.

29. Manap MA, Sulaiman WNA, Ramli MF, Pradhan B, Surip N (2013) A knowledge-driven GIS modeling technique for groundwater potential mapping at the Upper Langat Basin, Malaysia. Arab J Geosci 6(5):1621–1637

30. Mondal MdS, Pandey AC, Garg RD (2008) Groundwater prospects evaluation based on hydrogeomorphological mapping using high resolution satellite images: a case study in Uttarakhand. J Indian Soc Remote Sens 36(1):69–76

31. Nagaraju D, Papanna C, Siddalingamurthy S, Mahadevaswamy G, Nagesha PC, Rao K (2011) Identification of groundwater potential zones through remote sensing and GIS techniques in Kollegal Taluk, Chamarajnagar District, Karnataka, India. Int J Earth Sci Eng 4(4):651–658

32. Narendra K, Rao KN, Latha PS (2013) Integrating remote sensing and GIS for identification of groundwater prospective zones in the Narava Basin, Visakhapatnam Region, Andhra Pradesh. J Geol Soc India 81(2):248–260

33. Ndatuwong LG, Yadav GS (2014) Integration of hydrogeological factors for identification of groundwater potential zones using remote sensing and GIS techniques. J Geosci Geomat 2(1):11–16

34. Ozdemir A (2011) Using a binary logistic regression method and GIS for evaluating and mapping the groundwater spring potential in the Sultan Mountains (Aksehir, Turkey). J Hydrol 405(1–2):123–136

35. Pandian M, Kumanan CJ (2013) Geomatics approach to demarcate groundwater potential zones using remote sensing and GIS techniques in part of Trichy and Karur district, Tamilnadu, India. Arch Appl Sci Res 5(2):234–240

36. Pareta K, Pareta U (2011) Hydromorphogeological study of Karawan watershed using GIS and remote sensing techniques. Int Sci Res J 3(4):243–268

37. Pinder GF (2002) Groundwater modelling using geographical information systems. Wiley, New York, p 248

38. Pinto D, Shrestha S, Babel MS, Ninsawat S (2015) Delineation of groundwater potential zones in the Comoro watershed, Timor Leste using GIS, remote sensing and analytic hierarchy process (AHP) technique. Appl Water.

39. Prasad RK, Mondal NC, Banerjee P, Nandakumar MV, Singh VS (2008) Deciphering potential groundwater zone in hard rock through the application of GIS. Environ Geol 55(3):467–475

40. Preeja KR, Joseph S, Thomas J, Vijith H (2011) Identification of groundwater potential zones of a tropical river basin (Kerala, India) using remote sensing and GIS techniques. J Indian Soc Remote Sens 39(1):83–94

41. Rai B, Tiwari A, Dubey VS (2005) Identification of groundwater prospective zones by using remote sensing and geoelectrical methods in Jharia and Raniganj coalfields, Dhanbad district, Jharkhand state. J Earth Syst Sci 114(5):515–522

42. Rajaveni SP, Brindha K, Rajesh R, Elango L (2014) Spatial and temporal variation of groundwater level and its relation to drainage and intrusive rocks in a part of Nalgonda District, Andhra Pradesh, India. J Indian Soc Remote Sens 42(4):765–776

43. Rajesh R, Brindha K, Murugan R, Elango L (2012) Influence of hydrogeochemical processes on temporal changes in groundwater quality in a part of Nalgonda district, Andhra Pradesh, India. Environ Earth Sci 65:1203–1213

44. Ramaiah SN, Gopalakrishna GS, Vittala SS, Najeeb KMd (2012) Geomorphological mapping for identification of ground water potential zones in hard rock areas using geo-spatial information—a case study in Malur Taluk, Kolar District, Karnataka, India. Nat Environ Pollut Technol 11(3):369–376

45. Rao NS (2006) Groundwater potential index in a crystalline terrain using remote sensing data. Environ Geol 50(7):1067–1076

46. Rashid M, Lone MA, Ahmed S (2012) Integrating geospatial and ground geophysical information as guidelines for groundwater potential zones in hard rock terrains of south India. Environ Monit Assess 184(8):4829–4839

47. Sener E, Davraz A, Ozcelik M (2005) An integration of GIS and remote sensing in groundwater investigations: a case study in Burdur, Turkey. Hydrogeol J 13(5–6):826–834

48. Srinivasan K, Poongothai S, Chidambaram S (2013) Identification of groundwater potential zone by using GIS and electrical resistivity techniques in and around the Wellington reservoir, Cuddalore district, Tamilnadu, India. Eur Sci J 9(17):312–331

49. Suganthi S, Elango L, Subramanian SK (2013) Groundwater potential zonation by remote sensing and GIS techniques and its relation to the groundwater level in the coastal part of the Arani and Koratalai River Basin, Southern India. Earth Sci Res J 17(2):87–95

50. Varughese A, Suhail A, Chitra MG, Jiji PS, Deepthy C, Raneesh KY (2012) Identification of shallow groundwater potential zones using GIS—a case study. Int J Adv Eng Appl 1(4):65–70

51. Vittala SS, Govindaiah S, Gowda HH (2005) Evaluation of groundwater potential zones in the sub-watersheds of North Pennar river basin around Pavagada, Karnataka, India using remote sensing and GIS techniques. J Indian Soc Remote Sens 33(4):483–493

4

CHAPTER

Application of GIS and Remote Sensing to Groundwater Exploration in Al-Wala Basin in Jordan

Jawad T. Al-Bakri[1], Yahya Y. Al-Jahmany[2]

[1]*Department of Land, Water and Environment, Faculty of Agriculture, The University of Jordan, Amman, Jordan*

[2]*Department of Civil Engineering, College of Engineering, Majmaah University, Majmaah, Saudi Arabia*

ABSTRACT

Exploration of groundwater in countries with scarce water resources requires the implementation of effective tools that save time and money. In this study, geographic information systems (GIS) tools and remote sensing data were used to prepare and analyze digital layers of lithology, geological structure, drainage and topography to detect the most promising sites for groundwater exploration in an arid basin in Jordan. A separate map of existing wells was intersected with the generated maps to calculate the percentage of wells in each interval of density and count of lineaments and drainage. Different GIS functions of intersection and spatial query were then applied to produce the final map for the most promising sites for groundwater exploration. The possibility of using digital classification of remote sensing data for mapping the most promising sites for groundwater exploration was also investigated by applying unsupervised classification to a Landsat ETM+ image. Results showed that spatial distribution of the most promising sites for groundwater exploration was dependent on the interrelated factors of lithology, topography and geologic structure. The most promising sites were distributed within 4% of the study area. The highest percentage of groundwater wells was within the alluvial and wadi sediments, which were accurately detected by the digitally classified ETM+. The study showed that remote sensing and GIS provided efficient tools for mapping promising sites for groundwater exploration. However, the data of groundwater wells would contribute to refining the final locations of the most promising sites.

Keywords: GIS; Remote Sensing; Groundwater; Jordan

1. INTRODUCTION

Development of water resources represents a challenge for planners and decision-makers in countries with high population growth and water scarcity. In Jordan, the problem of water scarcity is being faced in all sectors and the current share of water is extremely low (145 m^3 per capita per year), as the country is dominated by an arid climate with low annual rainfall (**Figure 1**) [1]. Therefore, a considerable part of the agricultural production in the country depends on the available groundwater resources. The estimated safe annual yield from all renewable groundwater is 275 million cubic meters (MCM) while the annual withdrawal is 418 MCM [2]. The nonrenewable fossil aquifers are also providing 143 MCM per year for the coming 50 years [3].

At a population growth of 2.4%, an obvious increase in water demand is expected and the problem of water shortage will worsen with time [2,3], particularly under the adverse impacts of the expected climate change in the near future [3]. Without developing water resources, the expected annual water deficit will reach 488 and 769 MCM in years 2030 and 2050, respectively [3].

Currently, groundwater provides more than 45% of the total water resources in Jordan. The development of groundwater resources requires studies and surveys that identify areas with high potential for exploration to anticipate the inevitable geophysical surveys, which are expensive and time consuming. Groundwater exploration, however, has many problems such as the paucity of existing data and the high cost of data gathering [4]. Therefore, the integration of geologic and hydrogeological surveys with remote sensing data and geographic information systems (GIS) can contribute efficiently to these studies [4-6].

Remote sensing and GIS provide important data and tools for groundwater exploration [7-12]. Remotely sensed data of medium resolution are valuable sources

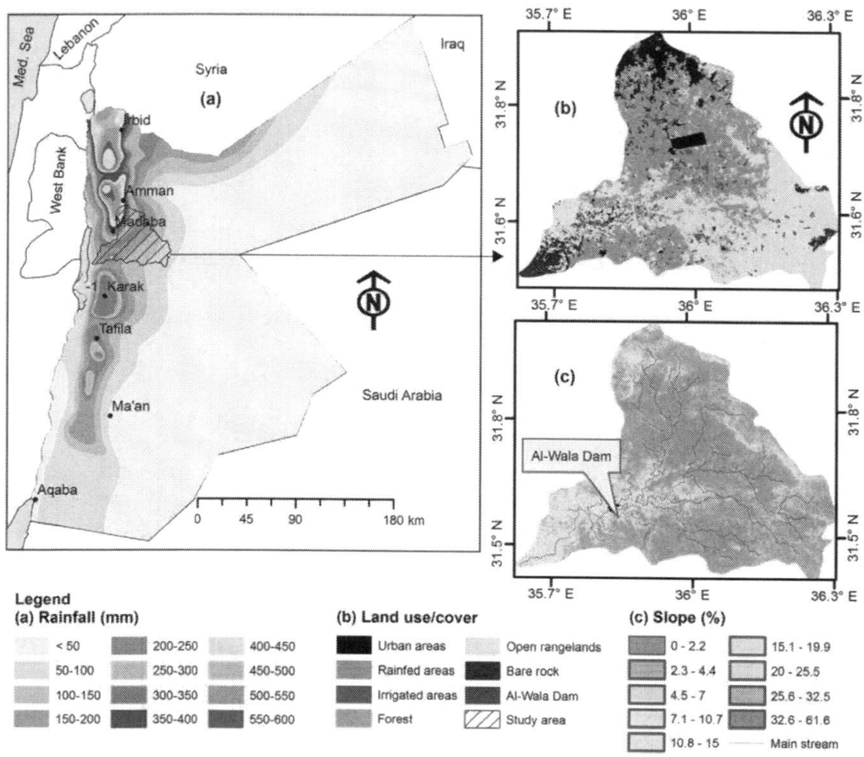

Figure 1. Location of the study area (a) and its land use/cover (b) and topography (c).

for creating maps of geological structures [4,13,14]. Maps of lithology, slope and drainage patterns can be also extracted from remote sensing data of aerial photography and medium to high resolution satellite imagery, depending on the scale of the study and the available resources. Conversion of maps into digital layers will enable the analysis of these maps by GIS tools. Implementation of spatial functions, query and intersection will produce maps of potential locations for groundwater exploration, based on different criteria that integrate maps of lineaments, topography, and lithology.

One approach for identifying potential sites for groundwater exploration is based on analyzing length density, frequency of lineament and drainage lines density to reflect different probabilities of groundwater potential in various parts of the area under consideration [7-9]. This approach was used in Jordan to identify promising sites with high potential for groundwater exploration [8,15]. Another approach for identifying these promising sites is based on analyzing hydro-geomorphological conditions to locate potential zones for groundwater recharge after applying a qualitative classification to the final map [11,14]. Reference [16] showed that this approach was useful in identifying areas for artificial recharge of groundwater, known as managed aquifer recharge (MAR).

In this study, remote sensing techniques and GIS tools are implemented to identify sites for groundwater exploration in Al-Wala basin in Jordan. The study aims to identify promising sites for groundwater exploration based on hydrogeological data. Also, it adds to the previous studies and efforts in Jordan by correlating locations with high potential with spectral classes derived from the digital classification of remotely sensed data of Landsat Enhanced Thematic Mapper Plus (ETM+). In terms of GIS approach, the study utilizes ground data to refine the selection of sites based on groundwater yield of the existing wells.

2. STUDY AREA

2.1. Location and Climate

The study covered Al-Wala basin in the middle parts of Jordan (**Figure 1**). The total area of the basin is 1900 km^2. The basin has a flat topography in the east and steep slopes in the west. Altitude of study area varies from 321 m below sea level at the eastern shoreline of the Dead Sea (lowest point on earth) to 997 m in the eastern parts of the study area.

The basin is dominated by an arid Mediterranean climate with relatively cold winters and hot summers. The mean monthly air temperature ranges from 6°C in January up to 22°C in August, with a mean annual air temperature of 15.0°C. With exception to northwestern parts, the mean annual rainfall is 100 - 200 mm, occurring mainly during winter season (October-April). The mean annual potential evaporation ranges from 1600 mm in the western parts to 2000 mm in the east and 2200 mm at the outlet near the Dead Sea. In terms of land use, about 47% of the basin is used as open rangelands. Agricultural activity is also taking place in the basin where rainfed cultivation of wheat and barley is practiced in 38% of the basin (**Figure 1**). Irrigated areas form 3% of the total area of the basin, while 7% of the basin is urban.

2.2. Geology

The basin is characterized by the strata of Belqa group and partly of Ajlun group (**Table 1**). The groups are mainly dominated by cretaceous carbonate rocks that

Table 1. Lithostratigraphic units and formations of the study area with percentage wells in each formation.

Group	Formation unit	Wells (%)
	Soil (S)	10.6
	Calcrete (Ct)	0.0
	Alluvial Gravels (Alg)	<0.1
	Alluvial Fans (Alf)	<0.1
	Alluvium and Wadi Sediments (Al)	18.9
	Travertine (Tr)	<0.1
	Fluviatile & Lacustrine Gravels (Pl)	25.6
Belqa	Basalt (B)	<0.1
	Wadi Shallalah Chalk (WSC)	<0.1
	Madaba Calcareous Breccia (MCB)	<0.1
	Umm Rijam Chert-Limestone (URC)	8.3
	Muwaqqar Chalk-Marl (MCM)	14.1
	Al Hisa Phosphorite (AHP)	11.1
	Amman Silicified Limestone (ASL)	1.8
	Wadi Umm Ghudran (WG)	6.7
	Wadi As Sir Limestone (WSL)	2.5
Ajlun	Fuhays/Hummar/Shuayb (F/H/S)	0.2
	Na'ur Limestone Formation (NL)	<0.1

outcrop in most of the study area [17]. The oldest rock that outcrops in the study area is the upper Wadi As Sir Limestone (WSL) formation, deposited in an agitated marine environment, from the Turonian age.

The lower part is composed of marl, marly limestone, coquina, micrite, sand and dolomicrite with chert nodules that has an average thickness of 62 m. The upper part is composed of limestone, dolomitic limestone and fossiloferous limestone with an average thickness of about 4.5 m [18].

An abrupt change in sedimentation resulting from an extensive Coniacian marine transgression marks the base of Belq group where a predominantly chalky unit of Wadi Umm Ghudran (WG) formation of ConiacianSantonian age overlies Ajlun group. This formation consists of Mujib Chalk, brecciated dolomitic limestone, and Dhiban Chalk with fish teeth outcropping all over the basin area. The formations of Belqa group include Al Hisa Phoshorite formation (AHP), (Campanian age), which consists of dark grayish chert, marly limestone and phosphate, and the Muwaqqar Chalk Marl formation (MCM) (Maastrichtian age), which consists of marl and chalk with chalky and micritic limestone concretions. Also, it includes Madaba Calcareous Breccia (MCB) (M. Eocene age), which consists of brecciated and partly conglomeratic clasts of phosphate, limestone, chert and chalk in a mudstone matrix located on hills, slopes and along some wadis in Madaba area.

The geology of the study area has some unique formations that include basaltic flows of Pleistocene, as volcanic tuffs restricted to the upper courses of Wadi Al Haidan. Quaternary Superficial Deposits (Pleistocene age) of Fluviatile and Lacustrine Gravels (PI) cover most of the wadis of the basin, particularly in the east. Aridic soils are dominating the low rainfall zones, while dark brown Mediterranean soils are dominating the high rainfall zones.

2.3. Hydrogeology

Groundwater of the basin flows from the eastern and northern highlands towards south and west till it reaches its downstream in the Dead Sea (**Figure 2**). Drainage pattern in the study area is modified by the direction of topographic displacement caused by the major faults so that it is directed towards faulted areas. The main wadis flow only in winter season and are directed ultimately westwards (**Figure 1**). The most important aquifers in the area are those within the upper cretaceous of AmmanWadi As Sir Limestone formations, with the intervening Wadi Umm Ghudran and overlaying Muwaqqar Chalk Marl formations, forming aquicludes [17,18].

The main source of groundwater is the aquifer system of Amman-Wadi As Sir formation known as B2/A7 aquifer system. Two recharge systems for this aquifer are recognized as follows:

1) Amman-Madaba, which is the main recharge system in the basin that includes the northwestern mountains in the basin. Groundwater of this recharge system flows southwards to Wadi Al-Wala and Al-Haidan (**Figure 2**).

2) Libb-Thiban, in which groundwater flows from the east and south towards Al-Haidan and through Al-Wala main stream.

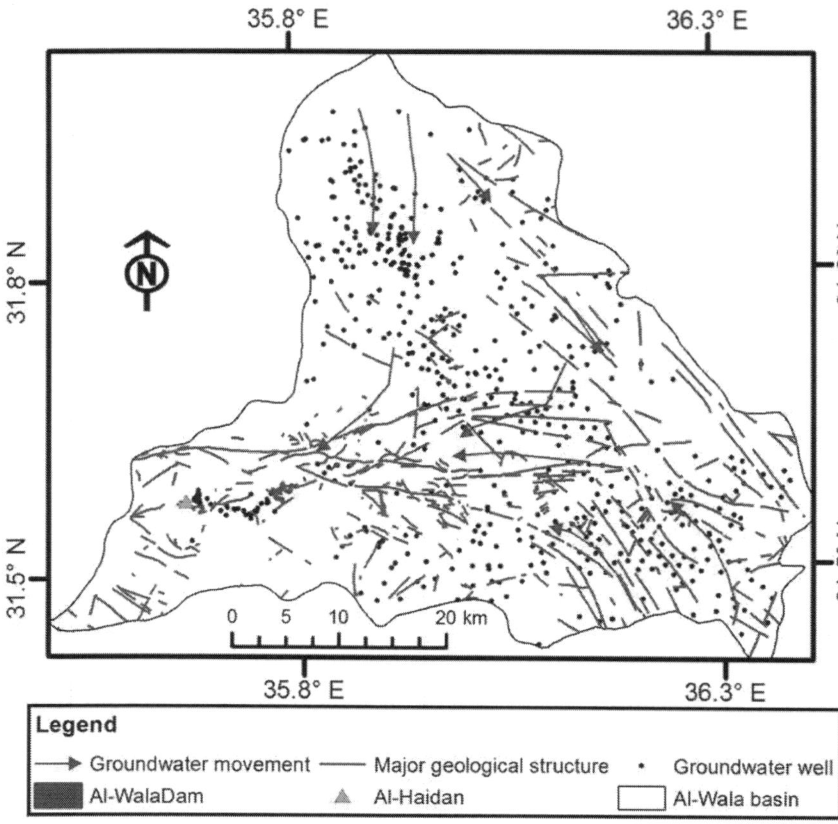

Figure 2. Major geological structures, groundwater wells and direction of groundwater movement in Al-Wala basin.

Ground water is the main source of water that is utilized for agriculture and municipal uses. The official records of Ministry of Water and Irrigation (MWI) for year 2010 show that the study area has 590 groundwater wells [19]. Fourteen springs are found in the downstream area in Al-Haidan. Water extracted from springs is used for both irrigation and domestic purposes.

Groundwater of the basin is highly important as water is pumped from Al-Haidan groundwater wells to supply drinking water to Madaba and southern parts of Amman. The MWI constructed Al-Wala dam, which became operational in the year 2003, to improve groundwater recharge. The current capacity of the dam is 9 MCM and the stored water is naturally and artificially recharging groundwater of Al-Haidan and other wells that are utilized for drinking and irrigation.

The future plan of MWI is to raise the height of the dam to store 26 MCM. The water will mainly be utilized for groundwater recharge in Al-Haidan and in the upstream area in the middle and the southern parts of the basin. Also, about 6 MCM will be utilized for irrigating fodder crops in the Jordan Valley [20]. Therefore, identifying the areas with high potential for groundwater exploration is of high importance for supporting agricultural activities in the area and for developing drinking water resources for the cities of Madaba and Amman.

3. METHODOLOGY

The methodology of this study was based on two approaches. The first was based on GIS, where layers of lithology, drainage patterns and lineaments were used to derive the most promising sites for groundwater exploration, as described by References [5,8]. Additionally, data on water quality and quantity were used to verify sites for groundwater exploration. The second part of the study was based on digital image processing (DIP) of remote sensing data to identify the spectral class that was highly correlated with the sites identified by the GIS tools. The first approach is based on the hypothesis that lineaments and drainage patterns would be the most important factors for evaluating the potential concentration of water in the fracture zone aquifer, while the second approach was proposed as a rapid method with minimum inputs for groundwater exploration. The following sections describe the methodology in more details.

3.1. Data Collection and Preparation

Hard copy maps of geology, topography and hydrology, covering the study area at a scale of 1:50,000, were acquired, scanned and geo-referenced in GIS. A satellite image of Landsat 7 ETM+ covering study area, except a small part in the east, was downloaded from the USGS website. The ETM+ imagery data which were used in this study included the visible (1, 2 and 3), the shortwave infrared (SWIR), the middle infrared (MIR) and the panchromatic bands. The spatial resolutions for these bands were 30 m for the visible to infrared and 15 m for the panchromatic. Geometric correction was carried out for the image using ground control points (GCP's) collected by a global positioning system (GPS). In total, fifteen points were used to correct the image using a second order transformation and the nearest neighbor resampling method [21].

A digital elevation model (DEM) at 30 m resolution, available from the data of the Advanced Spaceborne Thermal Emission and Reflection Radiometer (ASTER), was downloaded from the website of the provider (http:// asterweb.jpl.nasa.gov/gdem.asp). The DEM was used to generate a layer of streams network of the basin. The layer was overlaid over the geo-referenced topographic maps in GIS to correct the positional shifts. Line smoothing functions were also applied and manual editing was carried out for some of the streams which were either incorrectly delineated or had positional shifts. Data editing and improvement were also carried out for the layer of lineaments features (faults) using the panchromatic band of the ETM+, which was utilized to increase the details of this GIS layer.

Data on groundwater wells were obtained from the MWI. The data were transferred into spreadsheets and included the coordinates of wells, salinity of water and yield of well in cubic meters per hour (m^3/h). The yield was classified into three intervals following the criteria of the MWI for groundwater wells used in irrigation. The intervals were 35 m^3/h or less for low, 35 - 70 m^3/h for the medium and above 70 m^3/h for the high yield. The spreadsheet was transferred into a point layer in GIS and included the records for 560 wells.

3.2. Application of GIS

The output digital maps included layers of lineaments, stream (drainage) network and lithological formations with all attributes associated with each layer. **Figure 2** shows the lineament map, produced from this stage, and included all types of faults and strikes in the study area. The digital map of lineament was analyzed within GIS to identify types of structures in relation to ground water existence. Following this stage, all maps were rasterized to derive count (per·km^2) and length density (m/km^2) for both lineaments and stream network. Each of the output maps was reclassified into a new map with ten equal intervals of count and density.

The map of wells was intersected with the density maps of lineaments and drainage lines. The intersection results were then summarized in spreadsheets to calculate percentages of wells within each density interval of structure lines and drainage patterns. The same procedure of intersection was applied to determine the number and percentage of wells in each lithological unit. The intervals of count and density with maximum number of wells and the units of lithology were intersected to identify possible sites for groundwater exploration based on data of both medium and high yield wells. The output sites were based on the existence of 50% of wells or more within an interval or two for the medium and high yield groundwater wells. Wells with high levels of water salinity were excluded from the analysis. The final map was refined after considering altitudes, derived from the DEM, with maximum occurrence of medium and high yield of groundwater wells.

3.3. Application of Remote Sensing

The ETM+ imagery was transformed into three Principal Components (PC's), known as "Eigen Channels", which accounted for all variations of the selected bands of the ETM+. This was mainly performed to reduce data autocorrelation and dimensionality into fewer interpretable bands [21,22]. The unsupervised image classification was then applied to the output PC's using the Iterative Self-Organizing Data Analysis Technique (ISODATA) clustering algorithm to group the ETM+ grey levels into fewer classes, which were then identified in relation to lithological units using the existing geological maps and field visits.

The overall accuracy of the classification was assessed through a confusion matrix [21] by comparing a number of random points of classified pixels with existing geological maps and ground data. The output layer from the digital classification was intersected with the layer of the promising sites of groundwater wells to identify the class where most of these sites were located. The unsupervised classification was used as a rapid method that would automatically identify all possible spectral classes that corresponded to lithological formations in the study area. Therefore, results from DIP were intersected with results from GIS analysis of layers of lineaments and streams to assess the degree of agreement between the two methods.

4. RESULTS AND DISCUSSION

4.1. Lithology and Geological Structures

Results of intersecting the maps of wells and lithological formations are summarized in **Table 1**. Results showed that less than 26% of the wells were distributed in the fluviatile and lacustrine gravels, while about 19% of the wells in the study area existed in Alluvium and surrounding Wadi sediments (**Table 1**).

The intersection results also showed that about 76% of wells in the above mentioned formations were characterized by medium and high water yields. These results agreed with References [7,23] who found that most potentially excellent aquifers were found in valleys with thick sand or gravels covers. In Jordan, results of this study also confirmed the findings of References [8,15] who concluded that wells with high yield were found in thick coarse grained alluvial materials. These formations were expected to act as conduits for groundwater from deep to shallow aquifers. Therefore, most of the wells with good yield were found in these formations.

Analysis of lithological formation showed that structure features were observed above ground with considerable displacements found in Paleozoic formations. These were totally or partially absent in Mesozoic or Cretaceous lithological formations. This could be attributed to the stronger effect of stresses and faulting on older rock formations, as explained by Reference [5]. The stresses caused a deformation in the rock layers and produced geological structures that were identified and mapped as lineaments on the ETM+ imagery. Summary of the major fault in the study area is shown in**Table 2**.

In the study area, most sedimentary rocks had been deposited as horizontal strata. In various parts of the study area, subsequent uplifting and tilting caused the strata to be inclined. Furthermore, more than 70% of the main geologic structures were faults with tics showing downthrows side, faults inferred beneath superficial deposits and lineaments that were visible on the ETM+ imagery. All these structures were expected to form zones of high permeability layers which would recharge groundwater aquifers.

Table 2. Major fault groups in the study area.

Fault group name	Description
Wadi Zarqa Main Fault	Linear strike slip fault changes from a reverse to normal while moving from east to west.
Rumeil Fault	Normal fault extending for about 9 km with a downthrown of about 250 m near Wadi Rumeil area.
Wadi Al-Wala	Normal and reverse sets of faults with a downthrown to the south.
Az-Zafaran	A strike slip fault with dextral movement, starts with a reverse displacement and changes to a normal while moving from east to west. Displacement is more than 150 m near the desert highway where Muaqqar chalk marl is faulted against Amman Silicified Limestone formation.
Wadi Al Falij	Numerous small faults trending northeast-southwest in marble deposits.

4.2. Distribution of Wells within Count and Density Intervals

Results showed that wells with high yield were mainly located in Al-Haidan in the downstream area, while wells with low and medium yields were distributed all over the study area (**Figure 3**). Results of intersecting wells with counts and densities of lineaments and streams are presented in **Table 3**. In terms of lineaments count, wells were mainly found within the intervals of low count. Although this was unexpected, however it could be attributed to the fact that the lineaments count was already very low (less than 3.4 lineaments per km^2). On the other hand, the relationship between lineaments density and groundwater wells showed that 50% of groundwater wells were located within the lowest density interval. This could be explained by factor of topography, as many of the low and medium yield wells were located in flat areas in the northern part of the study area (**Figure 3**), where geological structures were not detected. In this part of the study area slope was less than 2% which would promote water infiltration and groundwater recharge [8,24]. In addition to topography, the high rainfall and the lithological formations (Holocene-recent alluvial and wadi sediments) enhanced groundwater recharge.

Although a considerable number of wells were found in the northern parts of the study area, however, most of the wells were characterized by low or medium yield. The high yield wells, on the other hand, were mainly within the moderate to high density of lineaments in the middle parts of the study area or in the low lineament density in the downstream area. In the middle parts of the study area, the high densities of lineaments were acting as points of groundwater recharge. In both locations, 69% of the wells were located in areas with an altitude of 650 - 750 m above sea level. None of the wells with high yield, on the other hand, occurred in altitudes of more than 750 m. This could be attributed to the direction of groundwater movement which was towards the southwestern parts of the study area (**Figure 2**). Similar findings of the impacts of topography on groundwater recharge were also indicated by References [11,15].

The combined impacts of topography, the high density of lineament and the alluvial deposits resulted in the presence of high yield wells in the middle parts of the study area around the main stream of Al-Wala. These results agreed with References [8,15] which indicated strong relationships between lineament density and locations of groundwater wells. In this study, none of the medium to high yield wells occurred in intervals that succeeded 5725 m/km^2 (Interval 6) (**Table 3**).

Querying groundwater wells with the map of drainage density showed that most of the groundwater wells were within the intervals of moderate drainage count and density (**Table 3**). The medium yield wells were mainly located within interval 5, while the high yield wells were mainly located within interval 4. Generally, drainage pattern is a reflection of the infiltration/precipitation ratio. This ratio is also controlled largely by permeability, which is in turn a function of the rock type and fracturing of the underlying rock or surface bedrock [7]. When comparing two terrain types, the one that contains the greatest drainage density is usually less permeable [7]. Therefore, this fact would imply that in groundwater exploration it would be important to combine the lineament map with drainage and lithological formations maps to identify the target sites of groundwater exploration, as explained in the next section. By this, variations in lithology, geologic history and response of bedrocks to stresses from folding to fracturing would be considered in groundwater exploration.

4.3. Intersection of Maps for Identifying Promising Sites

Results of intersecting the intervals of lineament and drainage count and density with the lithology and altitude are presented in **Figure 4**. The output of intersection presents the locations of promising sites for groundwater exploration. The map includes two classes that present medium and high probability sites based on the data of groundwater wells.

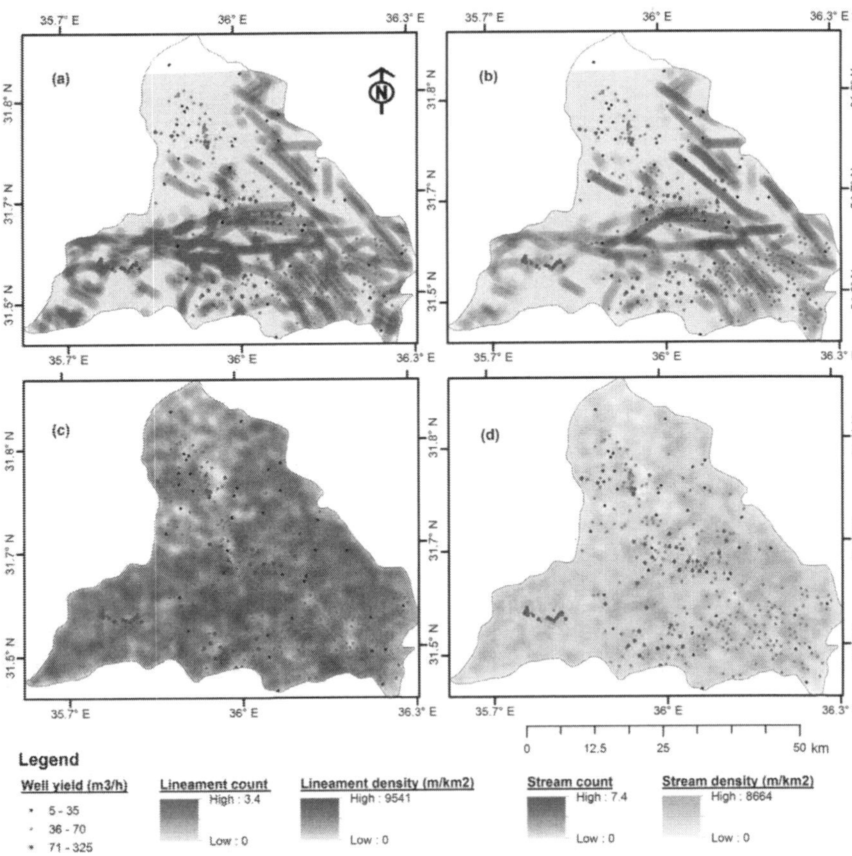

Figure 3. Distribution of groundwater wells within the layers of lineament count (a); lineament density (b); stream count (c) and stream density (d).

Table 3. Percent of wells within the different intervals of lineament and stream counts (per·km²) and densities (m/km²).

Interval	Lineament count	Wells (%)	Lineament density	Wells (%)	Stream count	Wells (%)	Stream density	Wells (%)
1	0.00 - 0.34	36.7	0 - 954	50.3	0.00 - 0.75	1.3	0 - 866	1.9
2	0.34 - 0.67	12.2	954 - 1908	20.2	0.76 - 1.49	4.3	867 - 1733	5.6
3	0.68 - 1.02	20.5	1909 - 2862	12.0	1.50 - 2.24	8.0	1734 - 2599	15.7
4	1.02 - 1.35	9.6	2863 - 3816	7.2	2.25 - 2.98	14.7	2600 - 3466	33.1
5	1.36 - 1.68	12.2	3817 - 4770	5.3	2.99 - 3.73	24.5	3467 - 4332	23.2
6	1.69 - 2.02	5.3	4771 - 5725	3.7	3.74 - 4.47	28.0	4333 - 5199	13.1
7	2.03 - 2.36	2.4	5726 - 6679	0.8	4.48 - 5.22	14.9	5200 - 6065	5.6
8	2.37 - 2.70	0.5	6680 - 7633	0.3	5.23 - 5.96	3.2	6066 - 6931	1.6
9	2.71 - 3.03	0.5	7634 - 8587	0.0	5.97 - 6.71	0.8	6932 - 7798	0.5
10	3.04 - 3.37	0.0	8588 - 9541	0.3	6.72 - 7.45	0.3	7799 - 8664	0.0

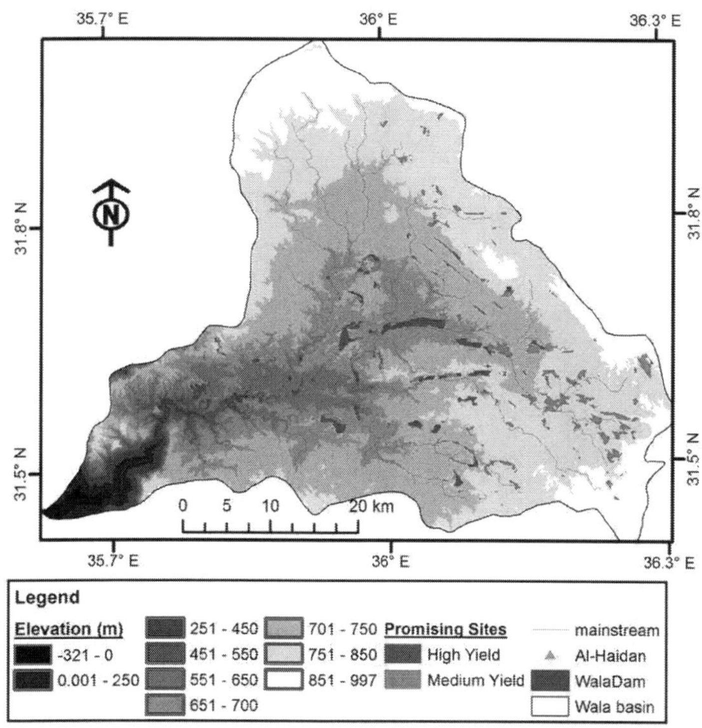

Figure 4. Locations of promising sites for groundwater exploration in Al-Wala based on the intersection of lineaments, lithology, drainage and altitude.

Results from this GIS approach showed that promising sites for groundwater exploration were distributed in different parts of the study area. However, most of the high yield sites were in the middle part of the study area on the main stream of Al-Wala. This part of the study area, mainly located within the formations of the fluviatile and lacustrine gravels (PI), was characterized by medium density of lineaments and streams, with altitudes ranging between 680 to 710 m. A small area of high yield was predicted in the southwestern part of the study area near Al-Wala dam. This area was classified as a high yield site due to its lower altitude and its occurrence on alluvial deposits (PI) with relatively moderate lineament densities (2863 - 3816 m/km^2).

The medium yield sites, on the other hand, were mainly distributed within the alluvium and wadi sediments (Al) and at altitudes range of 750 - 850 m. In terms of areas, the high yield sites had a total area of 26 km^2 while the medium yield sites had a total area of 50 km^2. These figures represent 1.4% and 2.6% of the study area for both high and medium yield sites, respectively.

4.4. Digital Classification of ETM+

Initial results of the unsupervised classification of the first three PC's, which accounted for all variations in the ETM+ image, produced a layer with ten spectral classes. The output layer (**Figure 5**) was compared with the map of lithological formations in the study area, and subsequently the final number of classes in the output layer was merged to 5, after considering class separation using

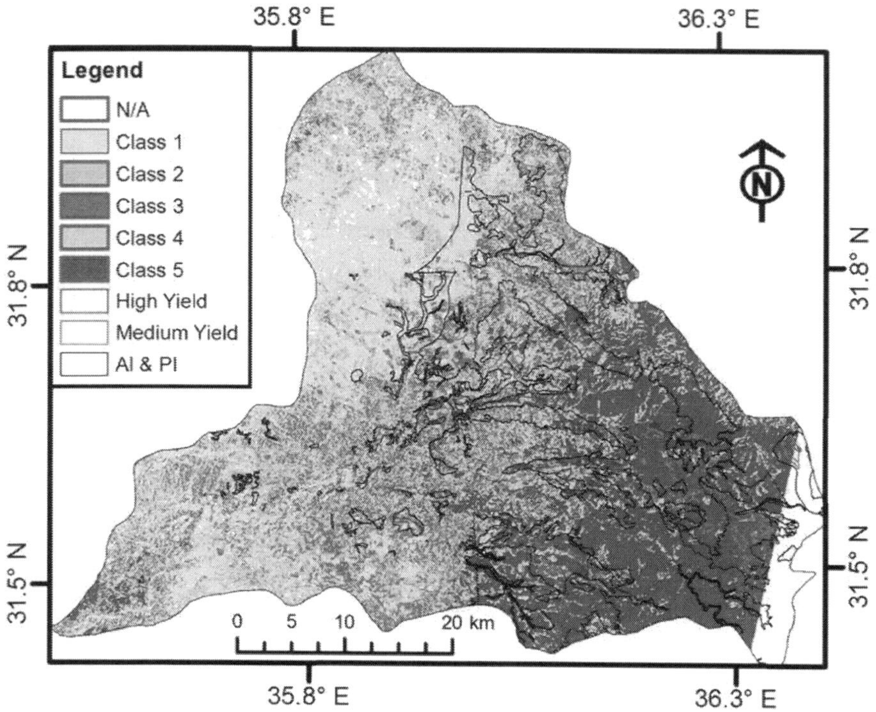

Figure 5. Distribution of alluvial formations and the most promising sites for groundwater exploration within the spectral classes of the unsupervised classification of the ETM+ image.

the method of transformed divergence [21]. **Table 4** summarizes the final classes and the major lithological units represented by each class. Classification accuracy varied from one class to another, with an overall accuracy of 76%. The variation in class accuracy was mainly attributed to the spectral mixing between classes and the accuracy of the original geologic maps.

Distribution of groundwater wells within the classified PC image (**Table 4**) showed that about 28% of groundwater wells were mainly located within class 5, representing alluvial formations and mapped with high accuracy. This figure was very close to results obtained from map intersection within GIS, which showed that 30% of groundwater wells were located within the alluvial formations.

Results showed that 71% and 52% of the promising sites for medium and high yield, respectively, were located within class 5. Also, 65% of the alluvial formations were located within this class. Therefore, outputs from digital classification would provide important thematic layers that could serve the purpose of identifying promising sites for groundwater exploration in basins similar to Al-Wala, particularly in locations where digital layers of geology might not be available. Other factors and data, however, should complement the DIP techniques. This was evident by the results of this study as the outputs from the unsupervised classification were not related to promising locations in the downstream area, where topography and lineament density were the main factors in identifying promising sites for groundwater exploration.

Table 4. Lithological description of the major classes produced from the unsupervised classification of principal components and the percentage of wells in each class.

Class	Description	Lithological Formations	Mapping Accuracy (%)	Wells+ (%)	Wells* (%)
1	Lithologic formations of late Cretaceous.	WG, ASL, AHP, MCM	73.8	29.8	33.7
2	Ajlun Group formations from Cenomanian to Turonian.	NL, F/H/S, WSL	68.9	7.5	2.7
3	Lithologic formations of the Pleistocene sediments of the geologic age, middle to upper Belqa Group.	PI, Tr	66.7	20.7	25.6
4	Lithologic formations of Tertiary (middle part of Belqa Group).	URC, MCB, WSC, B	70.8	14.1	8.3
5	Holocene-Recent alluvial and wadi sediments.	Al, Alf, Alg, S	95.0	27.9	29.5

+Based on intersection with thematic classes of unsupervised classification; *Based on intersection with lithological map (**Table 1**).

5. CONCLUSIONS AND RECOMMENDATIONS

Results from this study showed the significant roles of GIS tools and remote sensing data in identifying promising sites for groundwater exploration. The study provided an example that integrated real data on groundwater quantity and quality to refine the selection of the target sites. These data showed that sites with possible high yield were related to moderate intervals of lineament and stream densities. The other important data that affected the spatial distribution of the promising sites were lithology and topography. The study showed that GIS tools were very useful in refining the intervals of lineament and stream densities and in identifying the most promising sites based on data of existing groundwater well. Remote sensing data, on the other hand, could provide important information on geology and topography. Therefore, these tools and data sources would be recommended for groundwater exploration, as they would provide timely and cost effective tools for identifying and narrowing the target areas for groundwater exploration before carrying out further investigations.

6. ACKNOWLEDGEMENTS

This publication was supported by the NATO's Science for Peace Program, project SfP-983368 (2009-2012) "Assessment and monitoring of desertification in Jordan using remote sensing and bioindicators".

REFERENCES

1. "Annual Report of Jordan Meteorological Department," JMD (Jordan Meteorological Department), Amman, 2010

2. "Water for Life: Jordan's Water Strategy 2008-2022," MWI (Ministry of Water and Irrigation), Amman, 2009.

3. J. T. Al-Bakri, M. Salahat, A. Suleiman, M. Suifan, M. R. Hamdan, S. Khresat and T. Kandakji, "Impact of Climate and Land Use Changes on Water and Food Security in Jordan: Implications for Transcending 'The Tragedy of the Commons'," Sustainability, Vol. 5, No. 2, 2013, pp. 724-748.

4. A. Shaban, "Support of Space Techniques for Groundwater Exploration in Lebanon," Journal of Water Resource and Protection, Vol. 2 No. 5, 2010, pp. 469-477.

5. R. M. Teeuw, "Groundwater Exploration Using Remote Sensing and a Low-Cost Geographical Information System," Hydrogeology Journal, Vol. 3, No. 3, 1995, pp. 21- 30.

6. C. Travaglia and O. Ammar, "Groundwater Exploration by Satellite Remote Sensing in the Syrian Arab Republic," FAO Publications, Rome, 1998.

7. A. Edet, C. Okereke, S. Teme and E. Esu, "Application of Remote Sensing Data to Groundwater Exploration: A Case Study of the Cross River State, Southeastern Nigeria," Hydrogeology Journal, Vol. 6, No. 3, 1998, pp. 394-404.

8. A. El-Naqa, N. Hammouri, K. Ibrahim and M. El-Taj, "Integrated Approach for Groundwater Exploration in WadiAraba Using Remote Sensing and GIS," Jordan Journal of Civil Engineering, Vol. 3, No. 3, 2009, pp. 229-243.

9. I. A. Dar, K. Sankar and M. A. Dar, "Remote Sensing Technology and Geographic Information System Modeling: An Integrated Approach Towards the Mapping of Groundwater Potential Zones in Hardrock Terrain, Mamundiyar Basin," Journal of Hydrology, Vol. 394, No. 3-4, 2010, pp. 285-295.

10. M. Gupta and P. K. Srivastava, "Integrating GIS and Remote Sensing for Identification of Groundwater Potential Zones in the Hilly Terrain of Pavagarh, Gujarat, India," Water International, Vol. 35, No. 2, 2010, pp. 233-245.

11. N. Hammouri, A. El-Naqa and M. Barakat, "An Integrated Approach to Groundwater Exploration Using Remote Sensing and Geographic Information System," Journal of Water Resource and Protection, Vol. 4 No. 9, 2012, pp. 717-724.

12. S. Mukherjee, "Role of Satellite Sensors in Groundwater Exploration," Sensors, Vol. 8, No. 3, 2008, pp. 2006- 2016.

13. M. Yazdi, N. Sadati, A. A. Matkan and D. Ashoorloo, "Application of Remote Sensing in Monitoring of Faults," International Journal of Environmental Research, Vol. 5, No. 2, 2011, pp. 457-468.

14. S. Ganapuram, G. T. Vijaya Kumar, M. Krishna, E. Kahya and M. Cüneyd Demirel, "Mapping of Groundwater Potential Zones in the Musi Basin Using Remote Sensing Data and GIS," Advances in Engineering Software, Vol. 40, No. 7, 2009, pp. 506-518.

15. A. Al-Dababseh "An Integrated Approach to Groundwater Exploration of Northern Wadi Araba Basin Using Remote Sensing and Geographic Information System," MSc Thesis, The University of Jordan, Amman, 2003.

16. M. Alraggad and H. Jasem, "Managed Aquifer Recharge (MAR) through Surface Infiltration in the Azraq Basin/ Jordan," Journal of Water Resource and Protection, Vol. 2 No. 12, 2010, pp. 1057-1070.

17. NRA (Natural Resources Authority, Jordan), "Geological Map of Madaba, Sheet 3153 II, Scale 1:50000," National Geological Mapping Project 1988-1990, NRA, Amman, 1992.

18. N. Al-Hunjul, "The Geology of Madaba Area," Natural Resources Authority Press, Jordan, 1995.

19. MWI (Ministry of Water and Irrigation, Jordan), 2010.

20. MWI (Ministry of Water and Irrigation, Jordan), "Sustainable Fodder Crop Production," MWI, Jordan Valley Authority (JVA), Community Action Plan of the Badia Restoration Program (BRP), Ministry of Environment, Jordan, 2013.

21. J. R. Jensen, "Introductory Digital Image Processing: A Remote Sensing Perspective," 3rd Edition, Prentice Hall, Upper Saddle River, 2005.

22. R. A. Schowengerdt, "Remote Sensing: Models and Methods for Image Processing," 3rd Edition, Academic Press, London, 2006.

23. F. El-Baz, "Groundwater Concentration beneath Sand Fields in the Western Desert of Egypt: Indications by Radar Images from Space," Egyptian Journal of Remote Sensing and Space Sciences, Vol. 1, 1999, pp. 1-24.

24. N. S. Rao, "Groundwater Potential Index in a Crystalline Terrain Using Remote Sensing Data," Environmental Geology, Vol. 5, No. 7, 2006, pp. 1067-1076.

5

CHAPTER

Groundwater Vulnerability for the Surface Outcropping Aquifers in Jordan

Alsharifa Hind Mohammad[1], Thair Almomani[2], Ikhlas Alhejoj[1]

[1]Water, Energy and Environment Center, The University of Jordan, Amman, Jordan

[2]Ministry of Water and Irrigation, Amman, Jordan

ABSTRACT

Groundwater vulnerability is an overlay method that is used to determine the ability of pollutants to penetrate to the target aquifer and to harm it. This method helps decision makers by shedding light on pollution areas expected to pollute groundwater aquifers as caused by human activities on the ground surface. In Jordan, groundwater is the main water resource the country uses to match its demand. The groundwater basins in Jordan are divided into 12 major basins. Some basins are rechargeable and other basins are fossil. Many basins are over exploited. Amman Zarqa basin is a clear example for this case. Others are saline just like Azraq basin and the rest is expected to be affected by the growing demand for agricultural, municipal and industrial activities. In this study, a groundwater vulnerability map was produced for Jordan using DRASTIC index to study the vulnerability of the shallow aquifers throughout the country. The map shows different vulnerability classes ranging from low to very high reflecting the environmental, hydrological and hydrogeological settings of the groundwater and its recharge ability. The resulted map shows wide variation in groundwater vulnerability in different sites in Jordan. Areas with higher vulnerability are those with friable aquifer materials and shallow groundwater depths. Medium and low vulnerability classes are exist too because of the variations of the environmental settings within the targeted areas.

Keywords: Groundwater, Vulnerability, aquifers, vadose zone, Jordan

1. INTRODUCTION

Water supply is one of the main national concerns of Jordan that results from its limited water resources. The country was classified as one of the poorest countries in water resources worldwide [1] .

The concept of groundwater vulnerability was introduced in the 1960s in France by Margat (1968) [2] . Since then, several approaches for developing aquifer vulnerability maps were developed such as DRASTIC [3] , GOD [4] , AVI [5] , and SINTACS [6] . The conventional methods such as DRASTIC, AVI, GOD and SINTACS could distinguish degrees of vulnerability on a regional scale involving different lithologies [7] . DRASTIC was a familiar method developed for the US Environmental Protection Agency (EPA) by Aller et al. (1987) [3] and it had been used in different regions worldwide with satisfactory results [8] - [16] .

This article discusses the vulnerability of the shallow aquifers in Jordan using DRASTIC index; this index incorporates the different hydrological factors such as rainfall, depth to the groundwater table, hydraulic conductivity and other factors including topography, recharge ability and amounts, aquifer media and vadose zone effects which naturally participate in protecting aquifers.

Jordan is an arid to semi-arid country, with a land area of approximately 90,000 km^2. The mean annual rainfall in Jordan ranges between 50 mm in the southern desert region to about 600 mm along the western mountains adjacent to the Jordan Valley. As a result of increasing population pressure in recent years, Jordan has needed to increment exploitation of its natural resources as well as to protect the environment. Therefore, the demand for high quality drinking water is increasing, knowing that the average household water consumption is about 80 liters/capita/day which is one of the lowest rates worldwide compared to the living standards in Jordan [17] .

Groundwater is considered to be the major water supply source for many areas, and is the only water resource in some others. It can be divided into renewable and non-renewable groundwater resources. The main groundwater bodies are found in bedrock aquifers, and they form the main groundwater sources. The main aquifers are composed of sandstone like Kurnub and Ram groups, the carbonate aquifers like Amman Silicified Limestone A7/B2 and Belqa B4/B5 group, in addition to the basalt aquifer.

Unconsolidated aquifers, like fluvial deposits in the Jordan Valley, are compared to the bedrock aquifers of minor importance.

On a regional scale, the aquifers in Jordan can be grouped into three major aquifer systems, classified by their spatial distribution, lithology and age of the geological units.

These aquifer systems are:

1) Deep Aquifer Complex: it includes Ram aquifer, Khreim aquitard, Zarqa aquifer and Kurnub aquifer.

2) Intermediate Aquifer Complex: it includes A1/2 semi aquifer, A3 aquitard, A4 aquifer, A5/6 aquitard and B2/A7 aquifer.

3) Shallow Aquifer Complex: it includes B3 aquitard, B4/5 aquifer, basalt aquifer and alluvial deposits [18] .

2. METHODOLOGY

For building the groundwater vulnerability map, different environmental parameters interfere and participate to define the amount and location of impurities that may affect the aquifers such as; groundwater settings, hydrological and hydrogeological conditions, landuse parameters, environmental issues, soil parameters and other elements which may vary from one aquifer to other and from one area to another. These parameters are used to determine the vulnerability of groundwater [19] . The different spatial parameters of the aquifer (geology, recharge, water table, soil texture, etc.) are then exported into GIS, where equations for calculating groundwater vulnerability are used to deduce different vulnerability classes.

DRASTIC index for example is one of the vulnerability indices that could be applied in Jordan because of its applicability to Jordan's climate conditions, aquifer distribution and aquifers setting. Therefore, DRASTIC was chosen for this study In addition, DRASTIC index has wide variations of parameters that are really affecting the groundwater system in different environments. In this model (DRASTIC), spatial datasets on: Depth to groundwater, Recharge by rainfall, Aquifer type, Soil properties, Topography, Impact of the vadose zone and the hydraulic Conductivity of the aquifer are combined to assess the vulnerability of aquifers to surface activities (Table 1), [20] . The governing equation of the DRASTIC index DI was defined by Knox et al., (1993) [21] ; Fortin et al., (1997) [22] ; and Fritch et al., (2000) [23] ; is:

$$DI = DrDw + RrRw + ArAw + SrSw + TrTw + IrIw + CrCw \quad (1)$$

where DI is the DRASTIC Index, Dr is the rate of the D factor and Dw is the weight of the D factor, Rr is the rate for the Recharge factor and Rw is the weight for the Recharge factor, Ar is the rate for the Aquifer media factor and the Aw is the weight to the Aquifer media factor, Sr is the rate to the Soil media factor and Sw is the weight to this factor, the Tr is the rate to the Topography factor and the Tw is the weight to that factor, Ir is the rate of the impact of the vadose zone rate and Iw is the weight, and finally the Cr is the rate for the hydraulic Conductivity rate and the Cw is the weight to this factor; this DRASTIC index given in the equation above is considered as an indicator for pollution potential. The effects of the different parameters on the groundwater vulnerability were described by Piscopo (2001) [24] and are shortly described below.

Table 1. Parameter's classification for the use of DRASTIC index. [3] .

Parameter	Range	Rating	Relative weighting
Depth to water (D)	0 - 2 m	7	5
	2 - 5 m	6	
	5 - 9 m	5	
	9 - 15 m	4	
	15 - 23 m	3	
	23 - 30 m	2	
	>30 m	1	
recharge by rainfall	3	1	4
	4	2	
	5	3	
	6	4	
	7	5	
	8	6	
	9	7	
Aquifer media (A)	Massive shale	2	3
	Metamorphic/igneous	3	
	Weathered met./igneous	4	
	Bedded sandstone, limestone	6	
	Shale sequences	6	
	Massive sandstone	6	
	Massive limestone	6	
	Sand and gravel	8	
	Basalt	9	
	Karst limestone	10	
	Soil thin or absent	10	
	Gravel	9	
	Sand	8	
	Peat	7	

Soil media (S)	Shrinking and/or aggregated clay	4	2
	Sandy loam	5	
	Loam silty loam	4	
	Clay loam	3	
	Muck	2	
	Non-shrinking and non-aggregated clay	1	
Topography (T)	0% - 2%	7	1
	2% - 6 %	6	
	6% - 10%	5	
	10% - 16%	3	
	16% - 25%	2	
	>25%	1	
Impact of vadose zone	Confining layer	1	5
	Silt/clay	3	
	Shale	3	
	Limestone	6	
	Sandstone	6	
	Bedded limestone, sandstone shale	6	
	Sand and gravel with significant silt & clay	6	
	Metamorphic/igneous	4	
	Sand and gravel	8	
	Vesicular basalt	9	
	Karst limestone	10	
Hydraulic conductivity	0.50×10^{-6} - 0.50×10^{-4}	1	3
	0.50×10^{-4} - 0.15×10^{-3}	2	
	0.15×10^{-3} - 0.36×10^{-3}	4	
	0.36×10^{-3} - 0.51×10^{-3}	6	
	0.51×10^{-3} - 0.10×10^{-2}	8	
	$>0.10 \times 10^{-2}$	10	

For the first factor, Depth to water table (D); this factor is obtained for each outcropping aquifer using borehole data that contains information about the static water level from the monitoring of water wells available in the study area (Ministry of Water and Irrigation MWI). The D factor reflects the distance between the ground surface and the groundwater table; the depth of the water table reflects the safer condition to the pollution. For the depth to water table in Jordan, the resulting map is shown in Figure 1. The figure shows that along the western escarpments the shallower depth to water table is present which reflects the higher vulnerability values of these areas.

The second factor which is the (R) factor reflects the recharge rates and amounts in the targeted area. This factor is estimated as recharge from rainfall, which means that the recharge is affected by the amounts of the rainfall, the topography of the ground surface and finally the soil type and the porosity and permeability of the soils and rocks. Data from the Department of Meteorology (DOM in addition to the digital elevation model (DEM) to determine the slope of the land surface, and finally data from the Ministry of Agriculture to define soil types were used to find out the recharge factor in the area. In Jordan; the recharge values vary as shown in Figure 2 because of the wide variation in rainfall amounts in addition to the slope and soil components.

Geological maps from Natural Resources Authority, in addition to well logs and geologic cross sections were used to determine the third factor which is the aquifer media factor. The different aquifer types were considered and their outcrops over all Jordan were delineated in a map which shows the wide variations of this factor. The (A) factor map values ranged from 6 to 9 according to the type of the aquifer as shown in Figure 3, the value of 9 goes to the Bazalt aquifer type which is a high vulnerability rock type in the DRASTIC scale, then the lowest value 6 goes to the limestone aquifer across the study area .

Soil media (S), the fourth factor reflects the soils that cover the targeted areas, soils with high permeability permit the pollutants much easier than soils with low or no permeability that is reason for including this factor in this index. Soil media maps show a variation values from 2 to 10 depending on the soil types, the 10 reflects the high porosity and permeability soils which contains sand sized that and the lowest 2 value is for the soils that contain clay sized particles which lowering the ability of the water to be move through this soil type. Figure 4 shows the map for soil factor distribution.

The topography factor (T) depends on the slope setting of the surface of the ground which reflects the ability of pollutants to reach the groundwater according to the time needed for the water to infiltrate. Steep slopes are much safe than the gentle slopes. Figure 5 shows the wide variation of the topography in Jordan as slop percentages.

The Impact of the vadose zone factor (I) factor, this vulnerability factor is for the rocks above the aquifer, in this case where the article is about the outcropped aquifers which means that there are no layers over that targeted aquifers then the result is given to the higher I factor rate because of the absence of the protection layer here.

The last factor which is the hydraulic Conductivity factor (C) reflects the speed of the pollutants during its movement through the rock itself. Data are collected from MWI reports about transmisivity of aquifers. They were digitized in order to obtain the vulnerability.

3. RESULTS

Applying the equation of the weights and rates for DRATIC index for the natural setting in Jordan which in clude: depth to water table, aquifer type, recharge availability, soil material, topography, vadose zone and hydraulic conductivity four vulnerability classes were obtained. The Jordan Valley and Wadi Araba areas were excluded because of the sensitivity of the alluvium aquifer case such as the depth to water table then which requires detailed mapping and need detailed studies.

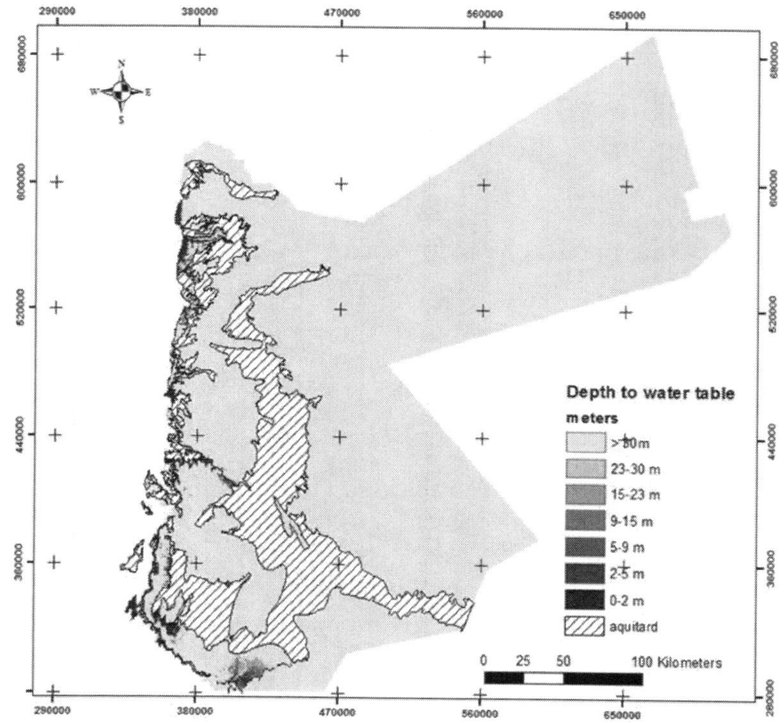

Figure 1. Depth to groundwater table (m) for the outcropping rocks in Jordan.

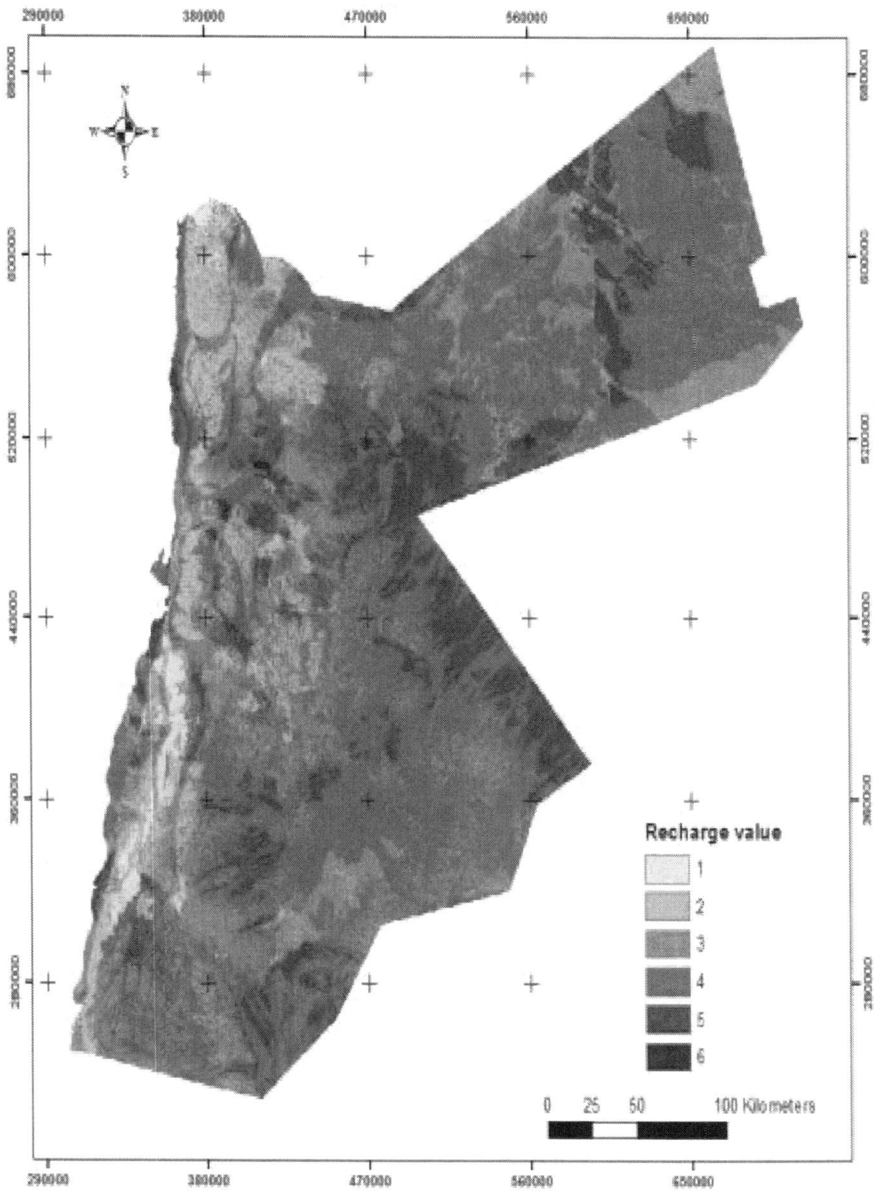

Figure 2. Recharge factor values for the outcropping rocks in Jordan.

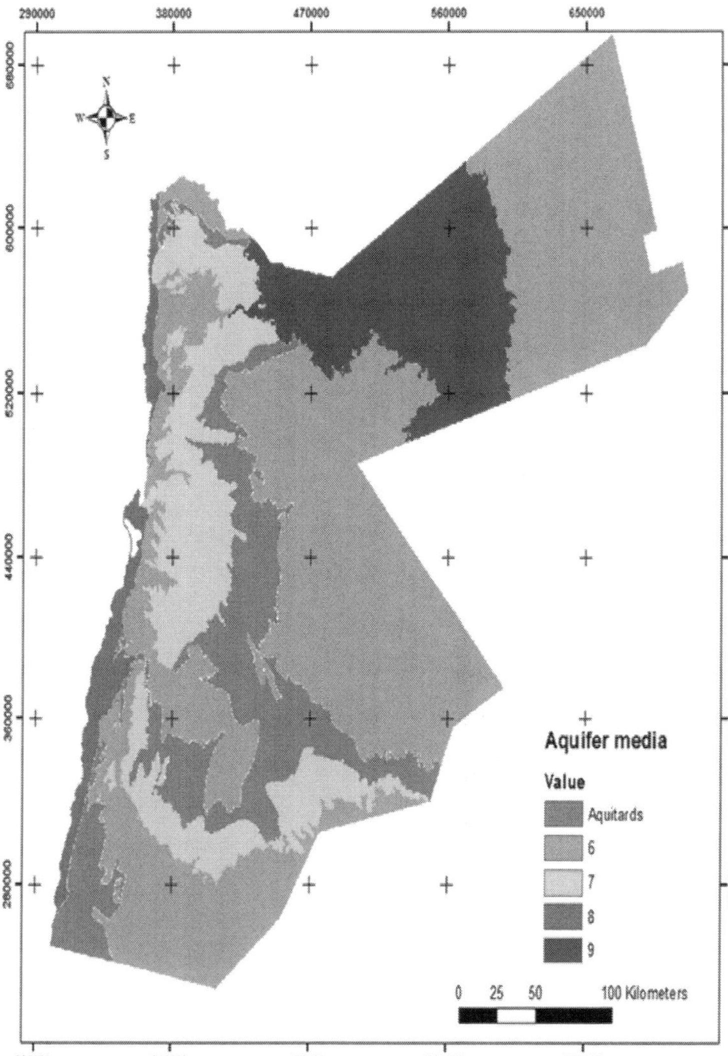

Figure 3. Aquifer media factor values for the outcropping rocks in Jordan.

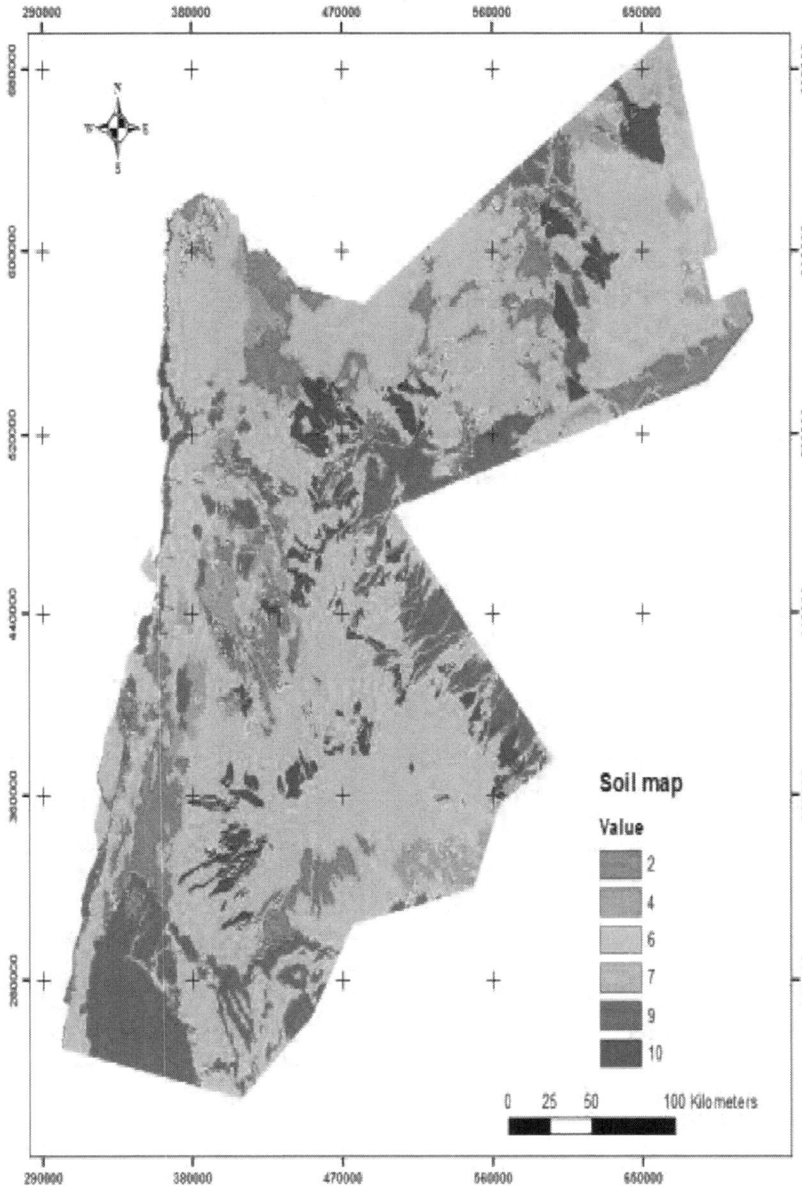

Figure 4. Soil factor values for the outcropping rocks in Jordan.

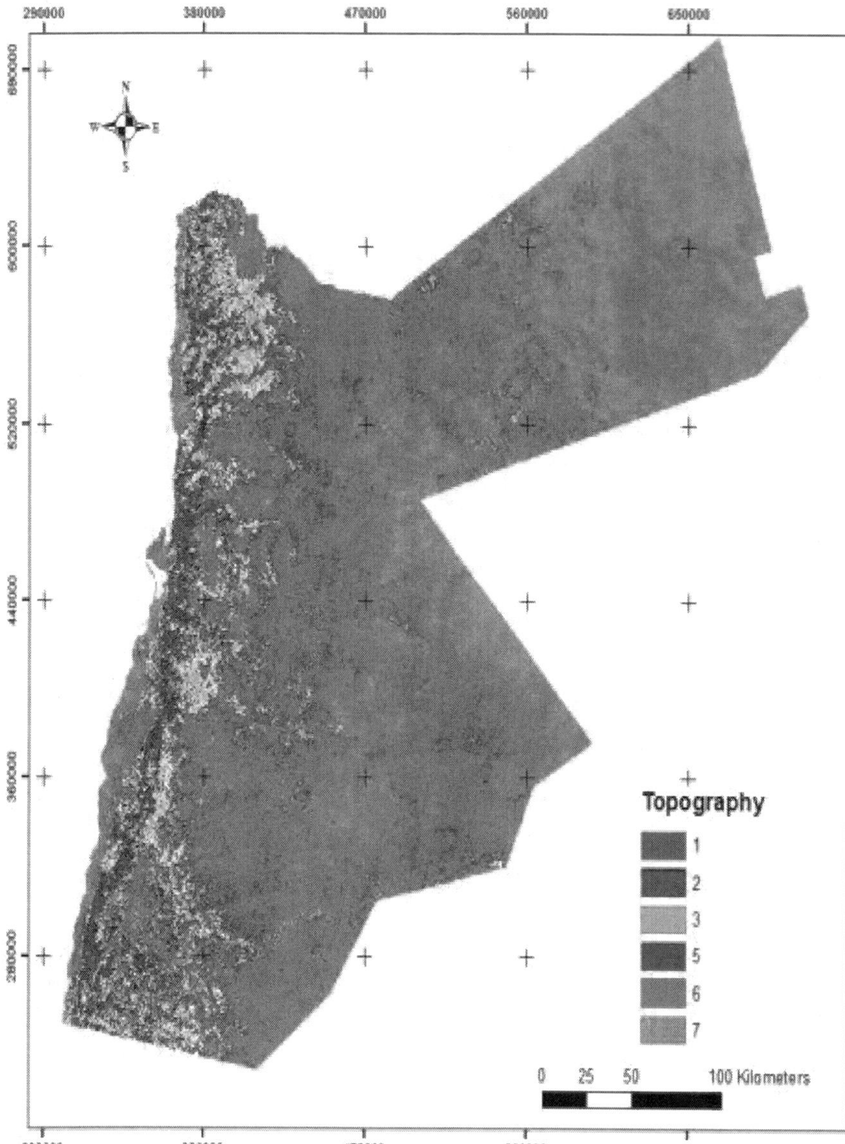

Figure 5. Topography factor values for the outcropping aquifers in Jordan.

Applying the process above as listed in Table 1, a vulnerability map can be produced that describes the ability of aquifers to be polluted as shown in Figure 6.

In this study, aquitards were excluded from the vulnerability map because this map refers to groundwater vulnerability and aquitards do not yield water. Figure 6 shows that the areas in central northern parts of the kingdom, added to some separated areas in the western parts are of higher vulnerability, this could be reasoned by the aquifer types and the soil classification in those areas, in addition to that; the depth to groundwater table in each aquifer which plays a vital role in assuming the vulnerability value.

Outcropping aquifers show a wide range of variation of vulnerability classes, ranging from very low vulnerability classes in the areas of very deep groundwater table and limited recharge up to high vulnerability classes in areas with shallow groundwater table and high potential of recharge. Figure 6 shows the variation of the ground- water vulnerability in Jordan and the variation of the vulnerability classes. Aquifer types within the study area play a major role for the vulnerability variation as shown is the alluvium areas are the higher vulnerability areas adding to that the gentle slope areas and the shallow water table.

Therefore, much care should be taken when planning projects of all types in the areas of high or medium vulnerability, whether these projects are industrial, agricultural or urban. Projects should be examined on their impacts on the groundwater resources, especially in

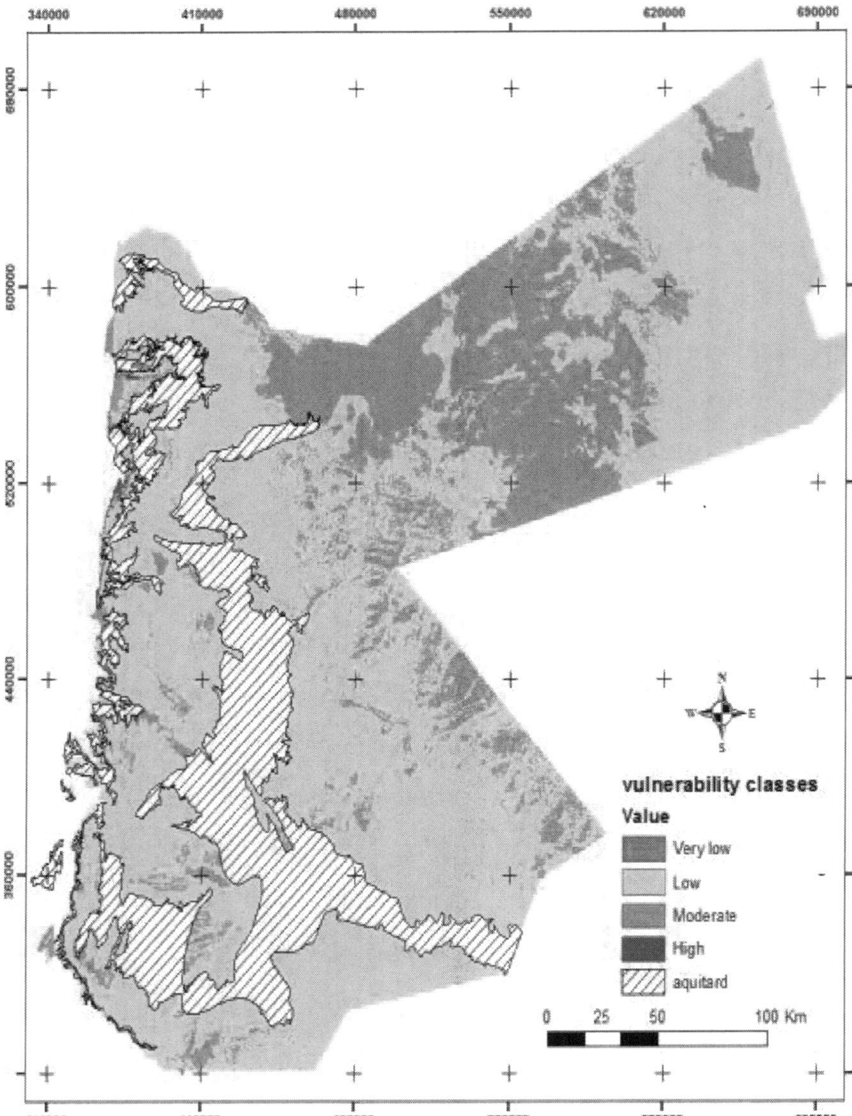

Figure 6. Groundwater vulnerability classes for the outcropping aquifers in Jordan.

terms of their pollution. Because, any damage of aquifers requires hundreds to thousands of years for restoration and rehabilitation.

For example, the area of Wadi As-Sir wastewater treatment plant is located in an area of medium vulnerability. This will lead to increasing potentiality of groundwater pollution in the area underlying the wastewater treat- ment plant and its surroundings.

Kherbit As-Samara wastewater treatment plan which is constructed in low vulnerability aquifer area but re-

charged in the area of the plant and along its effluent Wadi Dhulil causes very strong deteriorations in groundwater quality in that area.

Alsharifa Hind Mohammad, Thair Almomani, Ikhlas Alhejoj (irrigation return flows) and fertilizers contents.

REFERENCES

1. MWI (2002) Jordanian Ministry of Water and Irrigation. Open Files.
2. Margat, J. (1960) Carte Hydrogéologique du Basin de Fès-Meknès au 1/100,000. Agence du Bassin Hydrauliques de Sebou, Fès.
3. Aller, L., Bennet, T., Lehr, J.H., Petty, R.J. and Hachet, G. (1987) DRASTIC: A Standardised System for Evaluating Groundwater Pollution Potential Using Hydrogeologic Settings (EPA 600/2-87). Environmental Research Laboratory, Office of Research and Development, US Environmental Protection Agency Report, Tucson, 622.

4. Foster, S.S.D. (1987) Fundamental Concepts in Aquifer Vulnerability, Pollution Risk and Protection Strategy. In: van Duijvanbooden, W. and van Waegeningh, H.G., Eds., Vulnerability of Soil and Groundwater to Pollution, Proceedings and Information No. 38 of the International Conference Held in the Netherlands, in 1987, TNO Committee on Hydrological Research.

5. Van Stempvoort, D., Ewert, L. and Wassenaar, L. (1993) Aquifer Vulnerability Index (AVI): A GIS Compatible Method for Groundwater Vulnerability Mapping. Canadian Water Resources Journal, 18, 25-37. Giambastiani, B.M.S. (2007) Evoluzione Idrologica ed Idrogeologica Della Pineta di san Vitale (Ravenna). Ph.D. Thesis, Bologna University, Bologna.

6. Civita, M. (1994) Le carte della vulnerabilit`a degli acquiferi all'inquinamiento: Teoria e pratica [Contamination Vulnerability Mapping of the Aquifer: Theory and Practice]. Quaderni di Tecniche di Protezione Ambientale, Pitagora.

7. Vias, J.M., Andreo, B., Perles, M.J. and Carrasco, F. (2005) A Comparative Study of Four Schemes for Groundwater Vulnerability Mapping in a Diffuse Flow Carbonate Aquifer under Mediterranean Climatic Conditions. Environmental Geology, 47, 586-595.

8. Merchant, J.M. (1994) GIS-Based Groundwater Pollution Hazard Assessment: A Critical Review of the DRASTIC Model. Photograommetric Engineering & Remote Sensing, 60, 1117-1127.

9. Melloul, A.J. and Collin, M. (1998) A Proposed Index for Aquifer Water-Quality Assessment: The Case of Israel's Sharon Region. Journal of Environmental Management, 54, 131-142.

10. Cameron, E. and Peloso, GF. (2001) An Application of Fuzzy Logic to the Assessment of Aquifers' Pollution Potential. Environmental Geology, 40, 1305-1315.

11. Al-Adamat, R.A.N., Foster, I.D.L. and Baban, S.M.J. (2003) Groundwater Vulnerability and Risk Mapping for the Ba- saltic Aquifer of the Azraq Basin of Jordan Using GIS, Remote Sensing and DRASTIC. Applied Geography, 23, 303- 324.

12. Al-Farajat, M., Hamdan, I., Jaber, K. and Mohammed, S.H. (2005) GIS Mapping of Groundwater Vulnerability against Pollution in Amman Using DRASTIC Index. Hydrogeologie und Umwelt Heft, 33, 1-19.

13. Baalousha, H. (2006) Vulnerability Assessment for the Gaza Strip, Palestine Using DRASTIC. Environmental Geology, 50, 405-414.

14. Jamrah, A., Futaisi, A.A., Rajmohan, N. and Al-Yaroubi, S. (2007) Assessment of Groundwater Vulnerability in the Coastal Region of Oman Using DRASTIC Index Method in GIS Environment. Environmental Monitoring and Asses- sment, 147, 125-138.

15. Jasem, A. and Alraggad, M. (2010) Assessing Groundwater Vulnerability in Azraq Basin Area by a Modified DRASTIC Index. Journal of Water Resource and Protection, 2, 944-951.

16. Mohammad, A.H. (2014) New Groundwater Vulnerability Index for the Main Aquifers in Central Catchment Area in Jordan and Validation of the Results Using NO_3 Concentrations Maps. World Environment, 4, 22-32.

17. WAJ, Water Authority of Jordan (2012) Open Files.

18. (1977) National Water Master Plan of Jordan (NWMP). Vol. III, Surface Water Resources GTZ.

19. Vrba, J. and Zaporozec, A. (1994) Guidebook on Mapping Groundwater Vulnerability—IAH International Contributions to Hydrogeology, 16. FRG, Heise Publication, Hannover, 131 p.

20. Engel, B.A., Navulur, K.C.S., Cooper, B.S. and Hahn, L. (1996) Estimating Groundwater Vulnerability to Non-Point Source Pollution from Nitrates and Pesticides on a Regional Scale. In: Kovar, K. and Nachtnebel, H.P., Eds., Application of Geographic Information Systems in Hydrology and Water Resources Management, IAHS Press, IAHS Publication, Wallingford, 521-526.

21. Knox, R.C., Sabatini, D.A. and Canter, L.W. (1993) Subsurface Transport and Fate Processes. Lewis Publishing, Boca Raton.

22. Fortin, M., Thomson, K.P.B. and Edwards, G. (1997) The Role of Error Propagation for Integrating Multisource Data within Spatial Models: The Case of the DRASTIC Groundwater Vulnerability Model. Earth Surface Remote Sensing Procedure SPIE Conference, London, 358-361.

23. Fritch, T.G., McKnight, C.L., Yelderman Jr., J.C. and Arnold, J.G. (2000) An Aquifer Vulnerability Assessment of the Paluxy Aquifer, Central Texas, USA, Using GIS and a Modified DRASTIC Approach. Environmental Management, 25, 337-345.

24. Piscopo, G. (2001) Groundwater Vulnerability Map, Explanatory Notes. Castlereagh Catchment, NSW Department of Land and Water Conservation, Australia.

6

CHAPTER

Hydrogeochemical Evolution and Groundwater Flow in the Densu River Basin, Ghana

Dickson Adomako[1,2,3], Abass Gibrilla[2], Tetteh Thomas Akiti[1], Richmond Fianko[2], Piotr Maloszewski[3]

[1]*Graduate School of Nuclear and Allied Sciences, University of Ghana, Accra, Ghana*

[2]*Nuclear Chemistry and Environmental Research Centre, National Nuclear Research Institute, Ghana Atomic Energy Commission, Accra, Ghana*

[3]*Institute of Groundwater Ecology, German Research Centre for Environmental Health, Neuherberg, Germany*

ABSTRACT

Groundwater resources play a crucial role in the domestic and industrial water delivery system in Ghana. Hydrochemical and multivariate statistical techniques were used to investigate groundwater flow and geochemical evolution in the Densu River Basin aquifer system in Ghana. The hydrochemical and multivariate statistical techniques are mutually supportive and extracted information was analysed together with environmental isotope data. The results reveal three spatial groundwater bodies with defined hydrochemical facies, residence times, stable isotopic signals and hydrochemical evolution. The zones are designated as group one in the northern portion, group two in the transmission zone and group three as discharge in the southern portion of the catchment. Some of the stable isotope data of the groundwater do not lie close to the local meteoric water line (LMWL) on the δD-$\delta^{18}O$ plot and thus indicate that the meteoric water recharging the groundwater system has undergone some degree of evaporation.

Keywords: Densu, Groundwater, Oxygen-18, Deuterium

1. INTRODUCTION

Groundwater is one of the most important resources in Ghana. In many rural and urban communities in the Densu River basin, water supplies for domestic, irrigation and industrial uses are primarily dependent on existing groundwater resources. However, groundwater potential is under serious treat, due to increasing population density, mechanized agricultural practices, rapid urbanization, as well as domestic and industrial usage. Groundwater provision is time and again unsuccessful because of poor groundwater productivity of wells, drying of wells after prolonged draught or sometimes due to poor quality. The latter is hampered by lack of understanding the groundwater systems. However, groundwater resources in the Densu River basin have been fairly assessed and some data on the geology, borehole yield, static water levels and water quality analysis are available [1]. A large amount of hydrogeological and hydrochemical data have been generated through drilling activities in the basin. Nonetheless, due to lack of harmonized database, not much has been done to understand the hydrogeological and the hydrochemistry of the aquifers as a means of understanding the chemical evolution of groundwater system in the basin.

The changes in chemical composition of meteoric water recharging the aquifer will depend on factors such as soil-water interaction, residence time, congruent and incongruent dissolution of mineral assemblages, seawater intrusions and anthropogenic impacts. These factors would give rise to different water chemistry and hence different aquifer characteristics [2-5]. Additionally, in groundwater catchments where barrier boundaries and inputs are well defined, interpretation of the environmental isotopes record of precipitation, surface water and groundwater, coupled with available information on hydrochemical characteristics of groundwater will provide a detailed picture of the groundwater system [6,7].

Therefore, in other to understand the geochemical characteristics of the groundwater system, and its flow regime, multivariate statistical tools have been used widely to study the sources of the variation in the hydrochemistry of groundwater systems in recent studies. Arguably, one of the most powerful tools that have gained widespread acceptance in geo-environmental research community is multivariate statistical hydrochemical modelling. Precisely, factor and cluster analysis are the most widely used tools in geochemical studies to understand trends in data variation. Excellent examples of the application of factor analysis to solve hydrochemical problems are contained in [8-10]. Additionally, multivariate techniques have been used to resolve hydrological factors such as aquifer boundaries, groundwater flow paths and geochemical parameters e.g. [11,12]. It has also been used to identify geochemical controls on the composition of water [13], and separate anomalies such as anthropogenic impacts from background concentrations e.g. [14,15]. The specific objectives of this work are: 1) to understand the geochemistry and flow mechanisms of groundwater in the catchment, and 2) to characterize the isotopic ($\delta^{18}O$, δD) compositions of groundwater.

2. MATERIALS AND METHODS

2.1. Study Area

2.1.1. Location

The Densu River Basin is located in the southwestern side of Ghana. The Basin is characterized by undulating topography. Many of the hills here are craggy summits, which give a striking appearance to the landscape. The Densu River Basin lies between latitude 5°30'N - 6°20'N and longitude 0°10'W - 0°35'W (**Figure 1**). The Densu River Basin is coastal river catchments in the southern Ghana, and a typical rural community like all other rural settlements in Ghana. The Basin has a catchment area of 2,564 km^2 and spans 10 district assemblies and three regions of Ghana, namely Eastern, GreaterAccra and Central. There are about 200 settlements in the basin with a population density ranging from 150 to 200 per square kilometre, which is above the national average of 77 persons per square kilometre [16]. The river takes its source from the Atewa Range near Kibi in the East Akim Metropolitan Assembly, in the Eastern Region of Ghana and flows for 116 km from this source at an elevation of about 760 m (2,500 ft.) above sea level. The main tributaries of the river include Mame, Kuia, Adeiso, Dobro and Nsaki rivers. The river shares its catchment boundary with the Odaw and Volta basins to the east and north respectively, the Birim basin in the northwest and the Ayensu and Okrudu in the west. The river then enters the Gulf of Guinea through the tidal lagoon at Botianor, west of Accra.

2.1.2. Climate

The study area lies between two distinct climatic zones; the dry equatorial climate of the SE coastal plains, and the wet semi-equatorial climate further north from the coast. Both climatic zones are characterised by two rainfall seasons with different intensities [17]. The major rainy season extends from April/May to July and attains a peak in June when the maritime instability causes a surge of the moist south-westerly air stream resulting in the intensification of the monsoon rain. The minor season occurs between September and November when several disturbance lines give rise to local thunder activities, especially in the upper and middle belts which are covered with moist semi-deciduous forest. The total annual rainfall ranges from 1700 mm in the wet interior to 800 mm in the dry equatorial zone near the coast.

The climate of the study area is also characterised by prevailing high temperatures usually uniformly high throughout the year. The mean annual temperature is reported as 27°C, with March/April being the hottest (32°C) and August the coolest (23°C). The maximum and minimum relative humidity values vary from 89% to 93% and 41 to 72% respectively.

2.1.3. Geological Setting

The Densu River Basin is underlain mainly by the Proterozoic basin-type granitoid (the so-called Cape Coast batholiths), granite with associate gneiss. This geological formation covers a larger part of the basin (90%) as shown in **Figure 1**. The other formations are the Birimian formation, which occurs at the watershed of the river in the northwest, covering approximately 4% of the basin. The Togo Formation (6%) underlies the southeastern portion and the area south of Weija to the estuary portion of the basin. The Cape Coast granite Complex comprises heterogeneous group of rocks occupying most of the northwestern corner of the Akwapim North and part of Akwapim South districts. The bulk of the Cape Coast granite is composed of quartz, muscovite, orthoclase and albite. The granite changes from fine to medium grained, foliated biotite quartz diorite gneiss to hornblende–quartz diorite gneiss. These gneissic rocks are intruded by both acidic and basic igneous rocks, which include white and pink Pegmatite, aplites, granodiorite and peridotite.

Traditionally, [18,19] subdivided the Birimian Supergroup into two parts, the Lower Birimian, consisting mainly of dacitic/rhyodacitic volcaniclastics, greywackes, phyllites, argillites and chemical sediments [20], and the Upper Birimian, mainly made up of tholeiitic basalts with intercalated volcaniclastics, pyroclastic lava, and hypabyssal basic intrusive. Most of the rocks have been metamorphosed into calcareous chlorite schists and amphibolites (greenstones). Rocks belonging to the Birimian formation are sandstones, tuffs, grit, quartzites and some

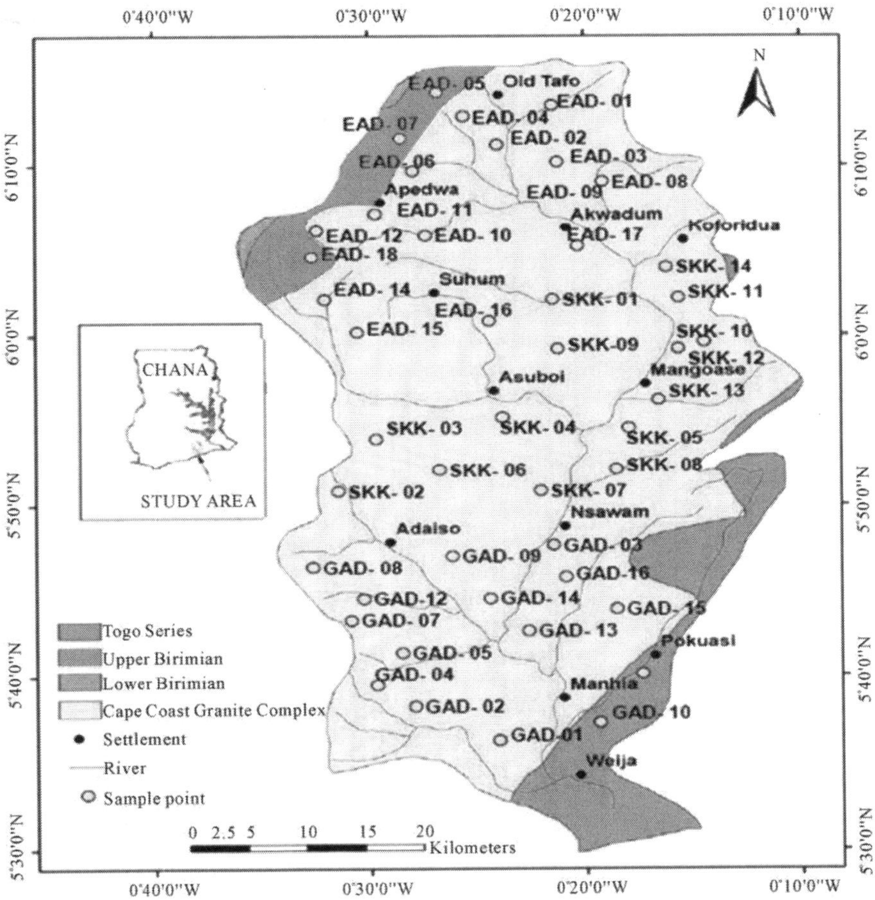

Figure 1. Map showing the sampling points in the study area.

breccia. The rocks of the Togo Series are highly folded and jointed, and form the chain of hills known as the Akwapim-Togo Ranges, which extend from the coast near Accra to the Togo border. The Togo Series consists of sandstone, quartzite and quartz schist, shale, phyllite and some talc mica schist.

2.1.4. Hydrogeology

The basement rocks underlying about 90% of the Densu River Basin have very little intergranular pore-space and are thus characterised by little or negligible primary porosity and permeability. Also, because of their limited transmissivity, crystalline basement aquifer exhibit comparatively low yields [21-24]. The two main features controlling groundwater occurrence in the basin are the secondary formed fractured and weathered zones respectively. The Birimian formation to the northwestern portion of the basin are generally well folded and deeply weathered, hence groundwater occurrence in this formation is very high. The granites in the basin are less fractured and weathered, hence groundwater occurrence are moderately low. The granitic aquifers thus formed are usually phreatic to semi-confined in character, structurally dependent and may be discontinuous in occurrence. However, in the northern portion of the granitic formation relatively high productive wells exist. The Togo series in the Southeastern portion of the basin is highly folded, jointed and fractured. The contact zone between the granites and the Togo formation provides a favourable condition for groundwater accumulation.

The depth of the borehole varies from 9.1 to 103.0 m in the granites, 23.0 to 40.0 m in the Birimian and 28.0 to 97.0 m in the Togo series [25]. The thickness of the mean weathered zone ranges from 1.0 and 32.0 m, 5.0 and 27.0 m and 3.0 and 36.0 m in the granites, Birimian and Togo series respectively. The static water level also ranges from 0.1 and 13.5 m, 0.8 and 16.9 m and 1.1 and 17.9 m in the granites, Birimian and Togo series respectively. Borehole yields are highly variable and lie in the range of 0.1 - 30.0 $m^3 \cdot h^{-1}$ with mean value of 2.0 $m^3 \cdot h^{-1}$ in the granite. In the Togo and the Birimian borehole yields are within 0.6 - 6.0 $m^3 \cdot h^{-1}$ and 0.7 - 9.0 $m^3 \cdot h^{-1}$ with the mean values of 2.8 $m^3 \cdot h^{-1}$ and 3.7 $m^3 \cdot h^{-1}$ respectively. The analytical results of groundwater chemistry and groundwater level data from the Basin show that groundwater is being recharged in the northern portion and flows to the southern portion of the basin. This is because the groundwater in the northern portion is characterized by low physio-chemical and chemical parameters, and these values increases along the flow line. Borehole yields in the basin however, vary from location to location; the water potential in the wells also varies from well to well [25].

2.1.5. Field Investigations

A total of 51 groundwater water samples were taken from the aquifers across the catchment during the period of August 2006, February and May-June 2007. The samples were collected from wells (**Figure 1**) at depth ranging between 24 and 65 m. Standard procedures with regard to well purging were followed for collection of the groundwater samples. The geographical location of all sampling sites was recorded using a hand-held global positioning system (GPS).

Alkalinity and physical parameters such as Electrical conductivity (EC), Total Dissolved Solids (TDS), temperature and pH of the samples were measured in the field using Hatch field titration kit and portable EC and pH meters. For groundwater samples, boreholes were purged until stable readings for the physical parameters were obtained. Samples were subsequently filtered through 0.45 um membranes and collected in acidwashed, well rinsed polyethylene bottles. Filtered and acidified (1% v/v HNO3) sampled were used for major cations, while filtered unacidified sampled were used for anion analysis. Unfiltered groundwater samples were collected in 30 ml glass bottles and tightly sealed for stable isotope analysis. Other samples were collected in 1 litre polyethylene bottles and preserved for tritium analysis.

2.1.6. Laboratory Analyses

Chemical and isotopic analysis of the water samples were performed at the Institute of Groundwater Ecology, German Research Center for Environmental Health (Neuherberg/Germany). Major cations (Ca^{2+}, Mg^{2+}, Na^{+}, and K^{+}) and anions (Cl^{-}, SO_4^{2-}, and NO_3^{-}) were analyzed using a Dionex DX 100 ion chromatograph equipped with a CS 12 and an AS14A-SC Ion Pac columns and an AS-40 auto-sampler. The accuracy of the duplicate analyses was estimated from the ionic balance error [26], which was within ±5% (3σ) for all samples.

Stable isotopes of oxygen and hydrogen were determined using isotope mass spectrometry. The water samples for oxygen isotopic analyses were prepared by conventional H_2O - CO_2 equilibration [27] where 5ml of each sample was equilibrated with CO_2 gas at 25 ± 0.1°C for 24 hours. The CO_2gas was then extracted and cryogenically purified in a vacuum line. For deuterium analysis, Cr at 850°C was used to produce hydrogen gas [28]. Both $\delta^{18}O$ and $\delta^{2}H$ values were determined relative to internal standards that were calibrated using Vienna-

Standard Mean Ocean Water (V-SMOW). The data were normalized following [29] as follows:

$\delta = [(R_S/R_{V\text{-}SMOW}) - 1] \times 1000$; where R_S represents either the $^{18}O/^{16}O$ or the $^{2}H/^{1}H$ ratio of the sample, and $R_{V\text{-}SMOW}$ is $^{18}O/^{16}O$ or the $^{2}H/^{1}H$ ratio of the V-SMOW. The analytical reproducibility is ±0.1% and ±1.0% for the oxygen and deuterium, respectively.

2.2. Multivariate Analysis

Multivariate analysis of the water quality data sets were performed using Cluster analysis (CA), Factor Analysis (FA) and Discriminant Analysis (DA) using Statistical Package for Social Sciences (SPSS 16.0) software package for windows.

2.2.1. Cluster Analysis

CA has been used by several researchers to classify a set of observed hydrochemical data within two or more mutually exclusive groups based on the combination of interval variables [10,30,31]. The Ward method [32] of hierarchical clustering is used for the factor scores. The Ward method applies clustering in such a way as to minimize the increment of the error sum of squares (IESS). The value of error sum of squares (ESS) for the i-th cluster is given by

$$ESS_i = \sum_{j=1}^{n_i}\sum_{k=1}^{p}\left(x_{ijk} - \overline{x_{ik}}\right)^2$$

where, x_{ijk} is the coordinate of the observation on the p variables for n_j, objects in the i-th cluster $\overline{x_{ik}}$ is the coordinate for the centroid in the i-th cluster, j is the number of samples, k is the number of variables. The total ESS for the total number of clusters g then can be obtained by summing ESS_i for i = 1 to g: $ESS = \sum_{i=1}^{g} ESS_i$.

When two clusters U and V are combined to form a new cluster, the IESS can be computed by

$$IESS = \left|\overline{x_u} - \overline{x_v}\right|^2 \Big/ \left(\frac{1}{n_u} + \frac{1}{n_v}\right)$$

where, $\overline{x_u}$ and $\overline{x_v}$ are the vectors for the centroid in the u-th cluster and in the v-th cluster, and n_u and n_v are the numbers of objects in U and V, respectively.

There are two modes of investigations namely R and Q-modes respectively. The R-mode is used to investigate the group associations or the affinity between the various variables in order to investigate their sources and processes with which they are associated [33]. The data set was further analyzed in Q-mode in order to identify similarities between cases. The similarity coefficient used was the simple distance defined in Euclidean space [34] and the clustering was performed by the average linkage (between-groups) method. The output results (water groups in the form of dendrogram) from the Hierarchical Cluster Analysis (HCA) are interpreted. Data treatment procedures and detailed explanation of the HCA clustering technique used by this study can be found in [30].

HCA was used to classify groundwater chemistry into logical groups that may be interpreted in terms of aquifer types, subsurface residence times and anthropogenic input on groundwater chemistry [13,35,36]. These techniques provides a good opportunity to identify relatively distinct paths for hydrochemical modeling and to understand geochemical evolution for a given aquifer system.

2.2.2. Factor Analysis

Factor analysis is a technique that purports to explain observed relations among several variates in terms of simpler relations that provide insight into the underlying structure of the variates. The factor analysis has also been used extensively in many geochemical studies to better understand the group associations or the affinity between the various variables that has been revealed by the cluster [8,34,37,38]. The simpler relations that are expressed in terms of a new set of variables are called factors.

In factor analysis, the observable random vector x, with p components, has mean μ and covariance matrix Σ. The factor model postulates that x is linearly dependent upon a few unobservable random variables F_1, F_2, F_3, $\cdots$, F_n, called common factors, and p additional sources of variation ε_1, ε_2, $\cdots$, ε_p, called errors or sometimes, specific factors. The factor analysis model is represented in matrix notion.

$$\chi - \mu = LF + \varepsilon$$

where, the matrix L is the matrix of factor loadings, the coefficient l_{ij} of L is called the loading of the i-th variable on the j-th factor, and F is the column vector components F_1, F_2, F_3, $\cdots$, F_m. When independence assumptions about the random vectors F and ε are used, the relation (Equation (5.13)) constitutes the orthogonal factor model.

To perform factor analysis, the first step is to compute a correlation matrix. This involves finding the correlation coefficient, which is a measure of interaction for all pairs of constituents. The second step in factor analysis is to estimate the factor loadings. The final step is to obtain easy interpretation of factors by factor rotation.

In this study, the principal component method is used as the parameter estimation method to transform a set of observed interdependent variables into an orthogonal set of variables called principal components. The first component accounts for as much as possible of the variances of the observed variates. The second principal component accounts for as much as possible of residual variance not accounted for by the first principal component. The third principal component also accounts for residual variance not accounted for by the second component.

Although numerous rotation schemes have been proposed, Kaiser's scheme called varimax rotation is used in this study. Varimax rotation yields a set of loadings such that the variance of the squares of the loadings is the maximum. Factor scores are commonly obtained by two approaches: the weighted least squares method and the regression method. The regression method [39] was used in this study to compute factor scores.

2.2.3. Discriminant Analysis

Discriminant analysis is a multicomponent statistical technique that has been used in many geochemical studies [8,34,37,38,40] to classify groundwater into mutually exclusive and exhaustive groups on the basis of a set of independent variables. The linear combinations of the independent variables found by means of this technique will discriminate the groups in such a way that the misclassification error rates are minimized. The objective was achieved by maximizing the between-group variance to the within-group variance [41].

To perform the discriminant analysis we need a qualitative dependent variable and a set of independent variables. The categorical-dependent variable is a grouping factor that allows each object to be placed in only one of the defined groups. In the linear discriminant equation;

$$D = B_0 + B_1 X_1 + B_2 X_2 + \cdots + B_p X_p,$$

where, $X_1, X_2, \cdots, X_p$ are the values of the independent variables, and $B_0, B_1, B_2, \cdots, B_p$ are the coefficients estimated from the data. The obtained function allows discriminating in the best way among the objects belonging to different groups; it therefore constitutes itself as a quality index, which integrates all the available information into a single variable [42].

3. RESULTS AND DISCUSSIONS

The groundwater pH is acid to circum-neutral and ranges from 5.3 and 7.2 with a mean value of 6.4. The temperatures are generally close to that of ambient temperature, 25.1 to 31.3°C, with a mean value of 27.8°C. The aquifer water presents wide variations in the total dissolved solids (TDS) and the electrical conductivity (EC) values; they ranges from 46 to 3890 mg/l and 99 to 8580 μS/cm, respectively and mean values are 601 mg/l and 1288 μS/cm. The high TDS and EC values of the groundwater samples in the study area follows the flow direction, thus from high to relatively low topographic areas towards the coast.

3.1. Statistical and Cluster Analysis

The SPSS 16.0 for windows statistical tool was utilized to support the conventional hydrochemical techniques. Twelve hydrochemical variables (EC, TDS, and pH, Temperature, Na^+, K^+, Ca^{2+}, Mg^{2+}, Cl^-, HCO_3^-, NO_3^- and SO_4^{2-}) were used in the statistical analyses. All the data were transformed to log base 10 values and subsequently standardized to their standard scores (z-cores) as outlined by [31]. For the R-mode, weighted pair-group method was applied and Euclidean distance was used as a measure of similarity. Before the analysis could be done, the data was normalised by generating a correlation coefficient matrix. The cluster revealed two distinct groups or clusters, and different clusters and their members were extracted as follows: cluster one includes EC, TDS, Na, SO_4, Cl, Mg, Ca, and K, while cluster two includes pH, HCO_3, temp, and NO_3. Furthermore, each variable measured via different units was rescaled to a range of 0 - 1 in order to have equal weight for the hierarchical cluster analysis (HCA). The result of the HCA of the hydrochemical classification is presented as a dendrogram in **Figure 2**. Based on the clusters, the groundwater samples in the study area were classified into three major groups. The mean water chemistry of the groups is given in **Table 1**, and their physical significance of the results is tested by relating the statistically defined clusters with their geographical locations. The most distinguishing factor between major groups seems to be TDS with concentrations increasing in both cations and anions, except NO_3^- in the following order:

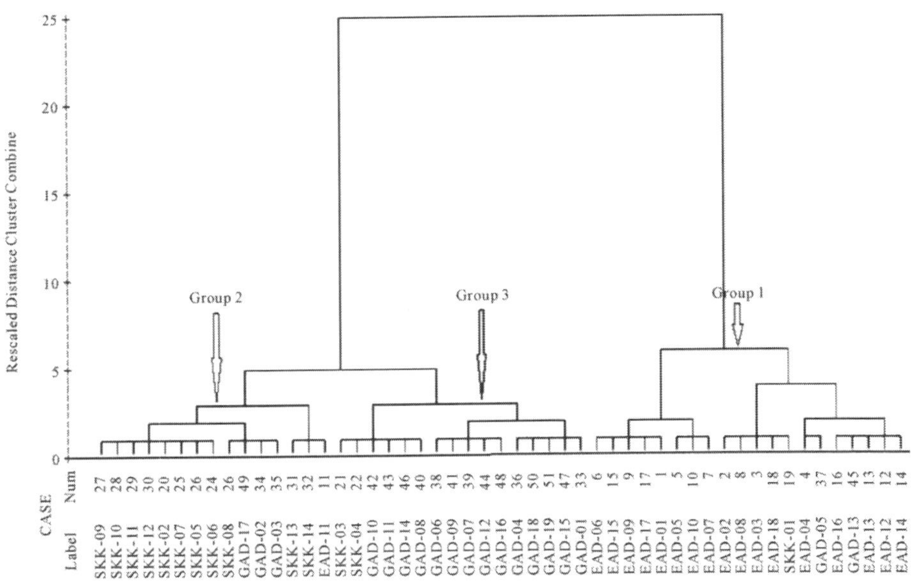

Figure 2. Dendrogram of HCA, including all water samples. The cluster analysis is based on parameters, Temp., TDS, EC, pH, Na^+, K^+, Mg^{2+}, Ca^{2+}, HCO_3^-, SO_4^{2-}, Cl^- and NO_3^-. The water samples are linked into clusters on the x-axis and the linkage distance is plotted on the y-axis.

3.2. Factor Analysis

The log-transformed data set was subjected to principal component analysis (PCA). The suitability of the data set for the analysis was assessed by Kaiser-Meyer-Oklin (KMO). The KMO value was 0.808, exceeding the recommended value of 0.6 [43]. Also, varimax rotation was performed to enhance group separation. The first three principal component extracted explain 80.86% of the variance in the data set. It is observed from **Table 2** that variables load fairly on only one component, thereby revealing a simple structure. Ca^{2+}, EC, TDS,

Table 1. Mean parameter composition and their corresponding standard deviation for the three groups of groundwater in the study area defined by HCA.

Parameters	Groups		
	Group 1 n = 14	Group 2 n = 21	Group 3 n = 16
Temperature (°C)	26.57 ± 0.74	28.21 ± 1.26	28.39 ± 0.84
pH	5.88 ± 0.30	6.50 ± 0.40	6.60 ± 0.26
Electrical conductivity (μS/cm)	348.50 ± 75.32	1219.02 ± 1739.44	2201.56 ± 1604.59
Total Dissolved Solids (TDS) (mg/l)	159.04 ± 75.32	574.31 ± 793.14	1023.75 ± 732.37
Ca^{2+} (mg/l)	24.44 ± 6.60	40.31 ± 20.55	57.53 ± 25.06
Mg^{2+} (mg/l)	3.28 ± 1.51	36.98 ± 23.54	64.37 ± 15.85
Na^{+} (mg/l)	16.25 ± 3.70	70.10 ± 39.48	84.44 ± 37.54
K^{+} (mg/l)	3.60 ± 2.10	5.26 ± 3.83	6.60 ± 4.54
HCO_3^- (mg/l)	65.42 ± 14.08	128.43 ± 49.90	210.33 ± 81.99
SO_4^{2-} (mg/l)	10.71 ± 5.80	65.11 ± 46.03	111.50 ± 81.99
Cl^- (mg/l)	18.68 ± 8.34	126.62 ± 111.67	223.68 ± 174.64
NO_3^- (mg/l)	9.93 ± 12.92	10.40 ± 14.26	2.58 ± 3.53

Table 2. Factor loading matrix for the data set after varimax rotation.

Variable (log-transformed)	Components		
	1	2	3
Temperature (°C)	0.216	0.158	**0.736**
pH	0.124	**0.909**	0.168
Electrical conductivity (μS/cm)	**0.899**	0.217	0.159
Total Dissolved Solids (TDS) (mg/l)	**0.898**	0.120	0.168
Ca^{2+} (mg/l)	**0.863**	0.144	–0.049
Mg^{2+} (mg/l)	**0.742**	0.528	–0.186
Na^{+} (mg/l)	**0.830**	0.397	0.136
K^{+} (mg/l)	**0.768**	0.055	0.111
HCO_3^- (mg/l)	0.372	**0.845**	–0.121
SO_4^{2-} (mg/l)	**0.854**	0.310	0.233
Cl^- (mg/l)	**0.928**	0.141	–0.367
NO_3^- (mg/l)	–0.003	–0.106	**0.614**

K^+, SO_4^{2-}, Na^+, Mg^{2+} and Cl^- are loaded in the first principal component, pH and HCO_3^-, are loaded in the second component, and temperature and NO_3^- are loaded in the third component. The 12 constituents are treated as 11 observable quantity variables with which the covariance matrix is composed for the multivariate analysis. **Table 3** shows the eigen values representing factors, the eigen value, difference between factors, and the proportion of total sample variance explained by the factors. The correlation matrix among the 12 physico-chemical variables of the groundwaters were presented in **Table 4**. A strong correlation was observed between EC and TDS, EC and SO_4^{2-}, EC and Cl^-, TDS and SO_4^{2-}, TDS and Cl^-, Na^+ and SO_4^{2-}, Na^+ and Cl^-, Na^+ and NO_3^-, SO_4^{2-} and Cl^-.

In Figures 3(a), (b) and (c), the major ion chemistry of groundwater in the study area for the various groups is summarized in three boxplots showing maximum, minimum, median, and the 25thand 75th percentile. The figure shows a wide range in concentrations for both plots, reflecting different water bodies. Group three groundwater samples have mean elevated concentration of 189.05 and 1009.50 mg/l for Cl^- and TDS. This may suggest dissolution of soluble salts.

3.3. Discriminant Analysis

Discriminant analysis was performed on the log-transformed data set to assess group validity for the groundwater samples in the study area. The sample ID was used as a basis for the groups for the samples. Mahalanobis distance-derived properties were used for group membership. Linear discrimination analysis (LDA) with cross-validation correctly classified 96.1% of the cases. **Figure 4**shows the groupings of the cases defined by the discriminant scores. Only two sampling points SSK-01 and GAD-11 were predicted to group one and three respectively, with their actual group being group two. In order to identify the variables that are most important in discriminating among the groups, step-wise discriminant analysis (SDA), with cross-validation were carried out with probabilities of F-to-enter-of 0.05 and F-to-remove of 0.10, 98.0% were classified by SDA. In all, a maximum number of 24 steps were employed. Temp., Ca^{2+},

Table 3. Eigenvalues of factors extracted through PCA, difference between factors, and proportion of variance explained by factors.

Factor	1	2	3	4	5	6	7	8	9	10	11
Eigenvalues	7.029	1.466	1.207	0.728	0.540	0.304	0.270	0.212	0.108	0.099	0.036
Difference	5.563	0.259	0.479	0.188	0.236	0.034	0.058	0.104	0.009	0.063	0.036
Proportion	0.592	0.121	0.089	0.061	0.045	0.034	0.237	0.164	0.081	0.055	0.025
Cumulative	0.592	0.713	0.801	0.862	0.908	0.942	0.964	0.984	0.992	0.997	1.000

Table 4. Correlation matrix among 12 physico-chemical variables of groundwater[a].

	T	pH	Ec	TDS	Ca^{2+}	Mg^{2+}	Na^{+}	K^{+}	HCO_3^-	SO_4^{2-}	Cl^-	NO_3^-
T	1.00											
pH	0.04	1.00										
Ec	0.37	–0.16	1.00									
TDS	0.37	–0.18	**0.99**	1.00								
Ca^{2+}	0.38	–0.18	0.40	0.40	1.00							
Mg^{2+}	–0.21	0.12	0.24	0.23	0.05	1.00						
Na^{+}	0.05	0.08	0.40	0.40	0.13	0.30	1.00					
K^{+}	0.05	–0.16	0.38	0.38	0.28	–0.06	0.26	1.00				
HCO_3^-	–0.15	0.41	–0.16	–0.17	–0.09	–0.02	0.22	–0.13	1.00			
SO_4^{2-}	0.33	–0.08	0.56	0.56	0.36	0.15	**0.60**	0.48	–0.13	1.00		
Cl^-	0.21	–0.05	0.60	**0.60**	0.40	0.45	**0.63**	0.30	0.32	**0.68**	1.00	
NO_3^-	0.29	0.11	0.48	0.48	0.13	0.06	**0.65**	0.14	0.09	0.48	0.44	1.00

[a]Factors greater than 11 not shown.

Mg^{2+}, HCO_3^-, Cl^- and NO_3^- were identify as the most discriminating variables for grouping the groundwater samples in the study area. About 90% of groups one, two and three of the groundwater samples in the study area are from the northern, central and southern portions of the study area respectively.

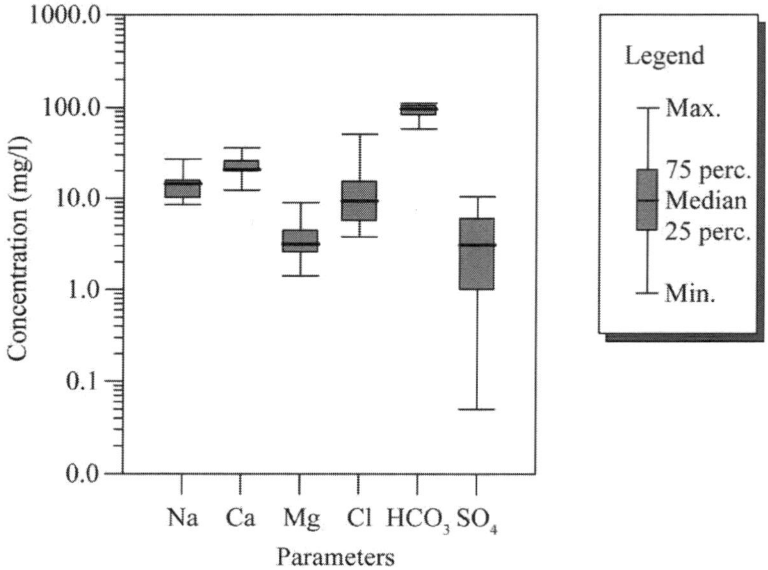
1000.0
100.0
10.0
1.0
0.1
0.0
Concentration (mg/l)
Na Ca Mg Cl HCO_3 SO_4
Parameters
Legend
Max.
75 perc.
Median
25 perc.
Min.

(a)

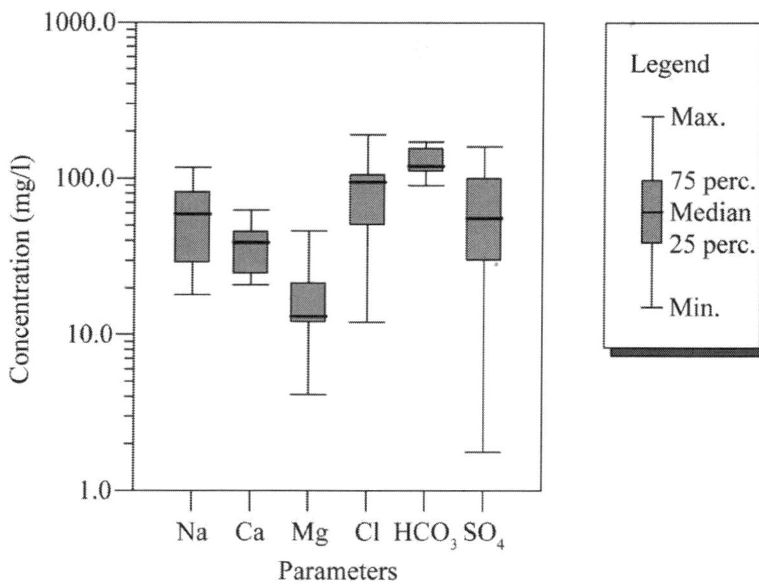
1000.0
100.0
10.0
1.0
Concentration (mg/l)
Na Ca Mg Cl HCO_3 SO_4
Parameters
Legend
Max.
75 perc.
Median
25 perc.
Min.

(b)

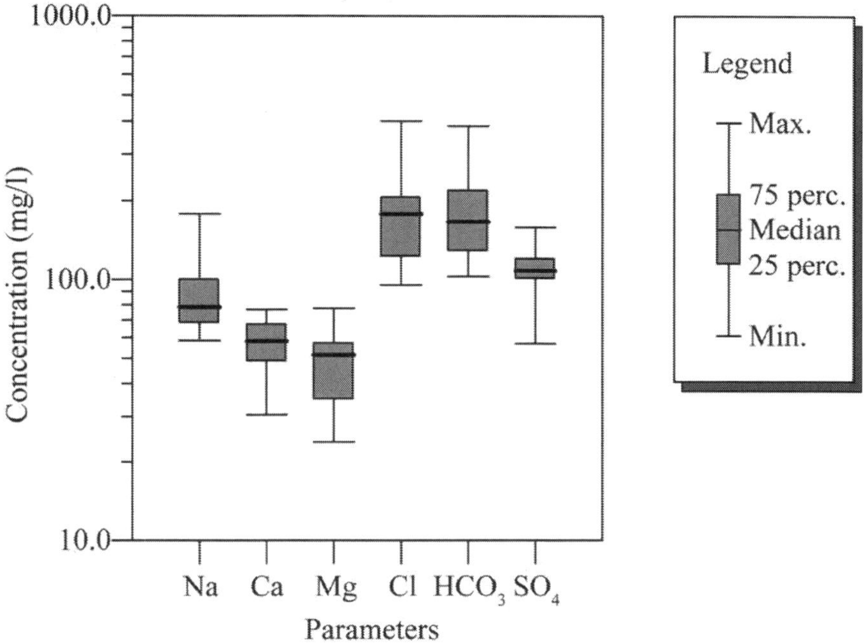

(c)

Figure 3. (a) Box plots of concentrations for major elements in groundwater samples from group one; (b) Box plots of concentrations for major elements in groundwater samples from group two; (c) Boxplots of concentrations for major elements in groundwater samples from groups one, two and three respectively.

3.4. Hydrochemical Facies

The chemical composition of groundwater samples in the Densu River basin is shown in **Figure 6**. The majority of the groundwater samples in group one of the study area are characterized by low TDS and Ca-HCO_3 water type (**Figure 5**). In most groundwater chemical evolution

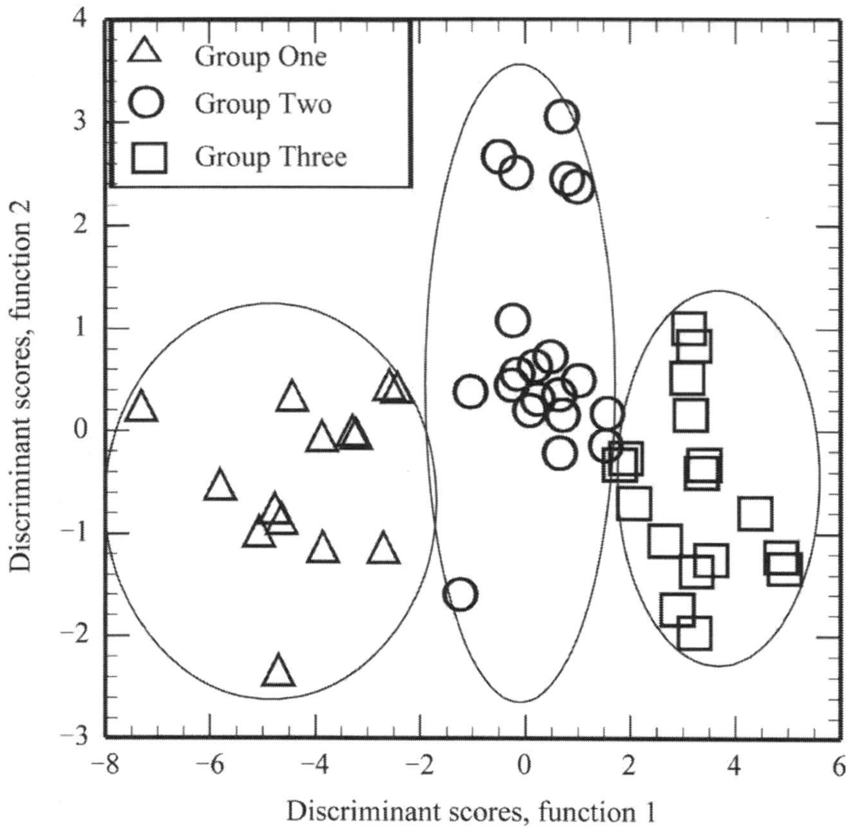

Figure 4. A plot of the first and second discriminant functions, showing groupings of the groundwater samples in the study area.

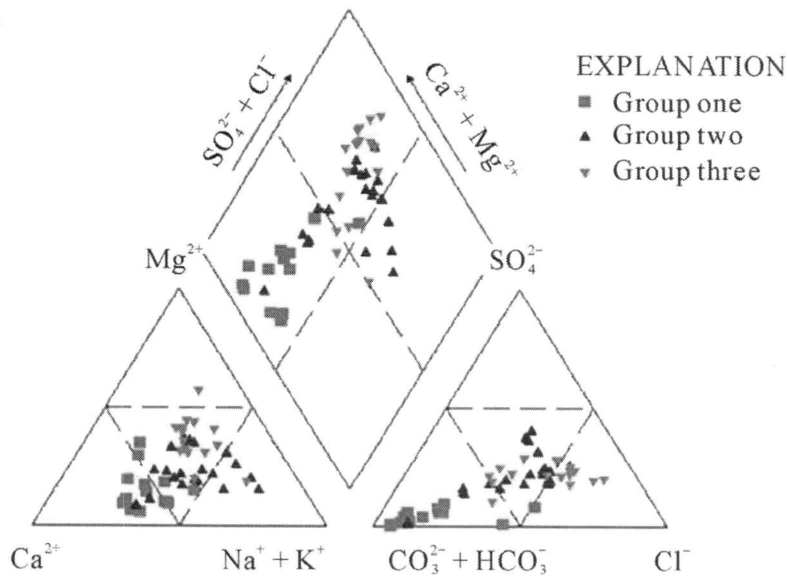

Figure 5. Piper plot of chemical data of groundwater samples in the study area.

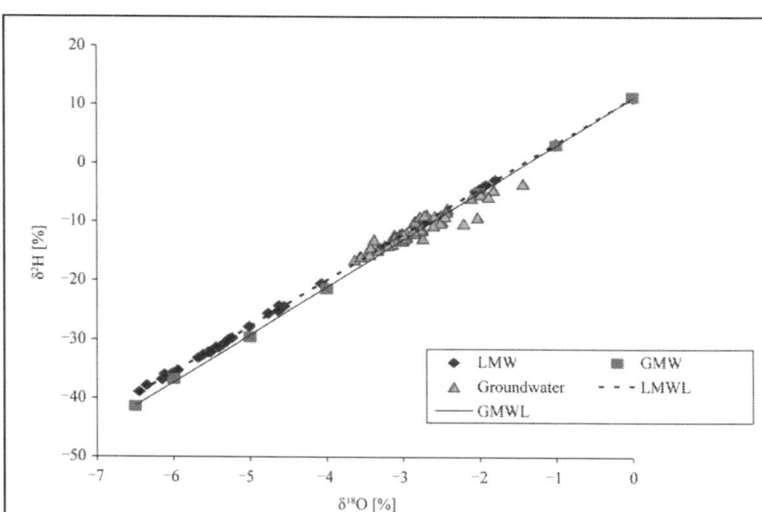

Figure 6. Regression line between Oxygen-18 and deuterium for the LMWL, GWML and groundwater.

model [44-46], this type is regarded as recharged water in areas where groundwater is at its early stage of geochemical evolution. This may imply that the rapidly circulating groundwater has not undergone pronounced mineral-water interactions. Groundwater from the recharge area is characterised by Mg^{2+}/Ca^{2+} ratio less than one, and is in favour of this argument. Groups two and three in the study area are derived from mixture of different water types ranging from Ca/Na-Cl, Ca/Na-HCO_3, and Na-Cl; this may be due to fractures contributing water from different places which may not flow in a well defined flow lines. This process is likely in crystalline aquifer context such as the Densu region. As groundwater proceeds along the flow line, incongruent dissolution may continue whereby the impurities in the solid phase are progressively released and a purer mineral (in this case Mg^{2+}) is produced.

Therefore, some of the groundwater samples will be characterized by Mg-HCO_3 water type. Most of the groundwater samples in groups two and three (**Figure 7**) show Na/Ca-HCO_3 water type, which may imply that the recharge waters might have undergone incongruent dissolution of silicate minerals, mainly plagioclase accompanied by cation exchange. The high TDS and the enrichment of Na^+ ion along the flow line testify that the groundwater has undergone relatively pronounced degree of chemical evolution. Also, Mg^{2+}/Ca^{2+} ratio greater than one for most of the groundwater samples in groups two and three confirm this process. This may imply long residence time of the groundwater samples in these groups.

3.5. Isotope Composition of Groundwater

The groundwater isotopic composition is an integrator of all recharge events; it will reflect the mixed isotopic signature roughly in proportion to the amount from flood recharge and "normal" recharge respectively [47,48]. The stable isotope contents appear fairly constant in time and they fingerprint the existence of different recharge episodes, if controlling factors such as temperature, humidity remain fairly stable. The isotopic content of groundwater in the study area ranges from –1.4% and –3.6% and –3.48% and –16.46% for $\delta^{18}O$ and δD, with a mean isotopic composition of –2.8% and –11.16% for $\delta^{18}O$ and δD respectively. A mean $\delta^{18}O$ isotopic composition of –2.8% was obtained by [49] for groundwater in the Accra plains of Ghana, which forms part of the study area. This may imply that the shallow aquifer system in the study area have not undergone significant degree of fractionation.

A regression of the groundwater data show them to lie on a gradient of 5.8, which slightly deviates from the gradient of 7.7 obtained for the Local Meteoric Water Line. This certainly indicates that the infiltrated rainwater might have undergone some degree of fractionation both on land surface and in the unsaturated (vadose) zone before entering the groundwater reservoir (**Figure 6**). In **Figure 7**, it is observed that groundwater samples with high chloride contents are relatively enriched. The probable cause may be congruent dissolution of dry chloride deposits in the soil zone, or mixing with seawater intrusion since, δD and $\delta^{18}O$ isotopic composition of seawater is zero. Similarly, in (**Figure 8**) there is a good correlation between $\delta^{18}O$ content and NO_3^- concentration in the groundwater samples. Few groundwater samples show elevated concentration of NO_3^- indicating anthropogenic point sources.

4. CONCLUSIONS

The Hierarchial Cluster Analysis (HCA) has classified the groundwaters of the study area into three groups. The advantage of the method was that the groups were objective, and clear geo-hydrological patterns were recognized. The subgroups show different degrees of water-rock interaction, residence time, as well as aquifer composition. The recharge zone, which is classified as group one samples are located in the northern portion of the basin. These aquifers are characterized by depleted δD and $\delta^{18}O$ values with low TDS. The transmission and the discharge zones are characterized by relatively enriched δD and $\delta^{18}O$ values, coupled with high TDS. This grouping was supported by the step-wise discriminant analysis (SDA) as 98% of the groundwater samples were grouped. The factor analysis also reveals that the variables were loaded fairly only in one component, thereby revealing a simple structure.

The dominant geochemical reactions in the basin are hydrolysis, incongruent dissolution, sulphate reduction as well as cation exchange. The recharge area is characterized by low TDS Ca-HCO_3 → Ca-Mg-HCO_3 water type to high TDS Na-HCO_3→NaCl water type in the discharge area, which is typical of groundwater flow mechanism as in [50]. Additionally, different water types in

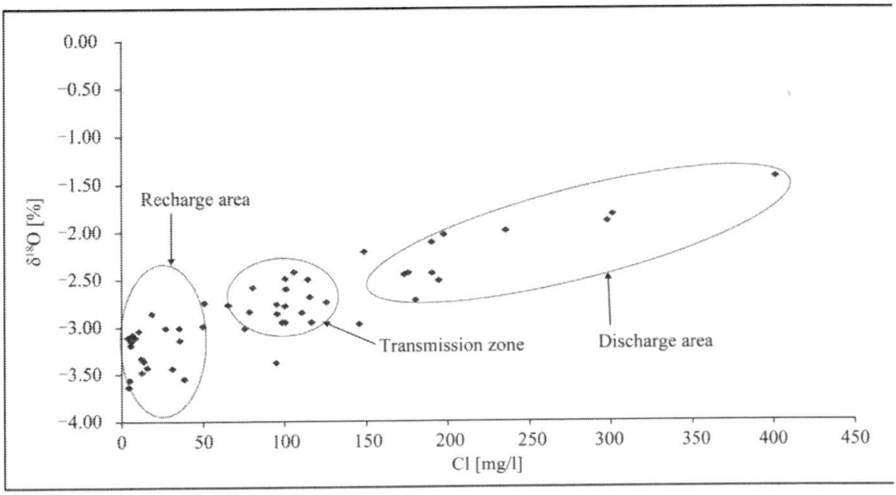

Figure 7. Scatter plot of oxygen-18 and chloride content of groundwater in the study area.

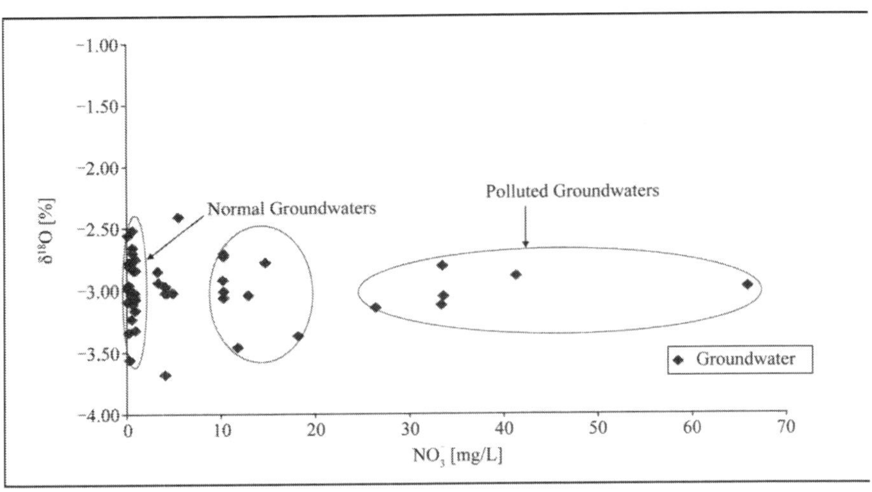

Figure 8. Scatter plot of oxygen-18 and nitrate content of groundwater in the study area.

the study area may be due to connected fractures in the weathered zone. Also, the Mg/Ca ratio in the groundwater samples increases along the flow line, as this may be used as an indicator of residence time and increasing maturity of groundwater. Generally, young groundwater will have low Mg concentrations as this was manifested in the groundwater samples in the study area. From the studies recharge to fractured aquifers by meteoric water may be through piston or preferential flow which has undergone some degree of fractionation. The aquifers seem to be connected as shown by the HCA and that the groundwater seems to flows from high topographic to lower topographic areas.

5. ACKNOWLEDGEMENTS

The authors are very grateful to the Director of National Nuclear Research Institute, Ghana Atomic Energy Commission and International Atomic Energy Agency for making funds available for this work. We also thank all the Technicians of the Chemistry Department, GAEC and Institute of Groundwater Ecology, Munich Germany for helping us during the sampling campaigns as well as the analysis.

REFERENCES

1. Water Research Institute, "Borehole Yield Map of Ghana," Water Research Institute, Accra, 1994.

2. S. Dapaah-Siakwan and P. Gyau-Boakye, "Hydro-chemical Framework and Borehole Yields in Ghana," Journal of Hydrology, Vol. 8, No. 4, 2000, pp. 845-852.

3. N. S. Rao, "Seasonal Variation of Groundwater Quality in a Part of Guntur District, Andhra Pradesh," Environmental Geology, Vol. 49, No. 3, 2006, pp. 413-429.

4. T. Helstrup, N. O. Jørgensen and B. Banoeng-Yakubo, "Investigation of Hydrochemical Characteristics of Groundwater from the Cretaceous-Eocene Limestone Aquifer in Southern Ghana and Southern Togo Using Hierarchical Cluster Analysis," Journal of Hydrology, Vol. 15, No. 5, 2007, pp. 977-989.

5. W. M. Edmunds and P. Shand, "Groundwater Baseline Quality," In: W. M. Edmunds and P. Shand, Eds., Natural Groundwater Quality, Blackwell Publishing Ltd., Malden, 2008, pp. 1-21.

6. C. Kendal, J. J. McDonnell, "Isotope Tracers in Catchment Hydrology," United State Geological Survey, Elsevier Science B.V., Amsterdam, 1998.

7. K. Didane, L. Bouchaou, Y. Hsissou and M. Krimissa, "Hydrochemical and Isotopic Characteristics of Groundwater in the Souss Upstream Basin, Southwestern Morocco," Journal of African Earth Sciences, Vol. 36, No. 4, 2003, pp. 315-327.

8. H. Suk and K. K. Lee, "Characterization of Hydrochemical System through Multivariate Analysis: Clustering into Zones," Groundwater, Vol. 37, No. 3, 1999, pp. 358-366.

9. F. Sanchez-Martos, R. Jimenez-Espinosa and A. Pulido- -Bosch, "Mapping Groundwater Quality Variables Using PCA and

Geostatistics: A Case Study of Bajo Andrax, Southeastern Spain," Hydrological Science, Vol. 46, No. 2, 2001, pp. 227-242.

10. M. Demile, S. Wohnlich, F. Wisotzky and B. Gizaw, "Groundwater Recharge, Flow and Hydrogeochemical Evolution in a Complex Volcanic Aquifer System, Central Ethiopia," Hydrogeology Journal, Vol. 15, No. 6, 2007, pp.1169-1181.

11. Y. Wang, T. Ma and Z. Luo, "Geostatistical and Geochemical Analysis of Surface Water Leakage into Groundwater on Regional Scale: A Case Study in the Liulin Karst System, Northwestern China," Hydrogeology Journal, Vol. 246, No. 1-4, 2001, pp. 223-234.

12. K. L. Locksey and M. E. Cox, "Statistical and Hydrochemical Methods to Compare Basalt and Basement Rock-Hosted Groundwaters: Atheron Tablelands, Northeastern Australia," Environmental Geology, Vol. 43, No. 6, 2003, pp. 698-713.

13. W. D. Alberto, D. M. del Pilar, A. M. Valeria, P. S. Fabiana, H. A. Cecilia and B. des los Angles, "Patterns Recognition Techniques for the Evaluation of Spatial and Temporal Variations in Water Quality, a Case Study: Suquia River Basin (Cordoba-Argentina)," Water Resources, Vol. 35, No. 12, 2001, pp. 2881-2894.

14. B. Helena, R. Pardo, M. Vega, E. Barrado, J. M. Fernandez and L. Fernandez, "Temporal Evolution of Groundwater Composition in an Alluvial Aquifer (Pissuerga River, Spain) by Principal

Component Analysis," Water Resources, Vol. 34, No. 3, 2000, pp. 807-816.

15. H. G. Pereira, S. Renca and J. Sataiva, "A Case Study on Geochemical Anomaly Identification through Principal Component Analysis Supplementary Projection," Applied Geochemistry, Vol. 18, No. 1, 2003, pp. 37-44.

16. N. Consult, "Rapid Environmental Assessment and Action Planning of Densu River Basin in Ghana," A Consulting Report, A United Nations Center for Human Settlements (Habitat) and Water Resources Commission of Ghana Funded, Accra, March 2001.

17. K. B. Dickson and G. Benneh, "A New Geography of Ghana," Longman Group UK Limited, Harlow, 1998, pp. 27-52.

18. N. R. Junner, "Gold in the Gold Coast," Gold Coast Geological Survey, Vol. 4, 1935, p. 67.

19. N. R. Junner, "Geology of the Gold Coast and Western Togoland. Geological," Survey Bulletin, Vol. 16, No. 11, 1940, pp. 5-10.

20. Leube and W. Hirdes, "The Birimian Supergroup of Ghana—Depositional Environment, Structural Development and Conceptual Model of an Early Proterozoic Suite," Federal Institute for Geosciences and Natural Resources, unpublished report 99529, Hanover, 1986, p. 260.

21. H. E. Gill, "A Groundwater Reconnaissance of the Republic of Ghana, with a Description of Geo Hydrologic Provinces,"

Geological Survey Water-Supply Paper 1757- -K, Washington, 1969.

22. Water Research Institute, "Groundwater Resources Assessment of Ghana," Water Research Institute/Council for Scientific and Industrial Research, Accra, 1999.

23. P. K. Darko and J. Krásný, "Regional Transmissivity Distribution and Groundwater Potential in Hard Rocks of Ghana," In: J. Krásný, Z. Hrkal and J. Bruthans, Eds., Groundwater in Fractured Rocks, Taylor & Francis, Boca Raton, 2003, pp. 45-46.

24. P. G. Cook, "A Guide to Regional Groundwater Flow in Fractured Rock Aquifers," Seaview Press, Henley Beach, 2003.

25. Water Resources Research Institute, "Groundwater Assessment: An element of integral Water Resources Management-The Case of the Densu River Basin," 2003.

26. R. A. Freeze and J. A. Cherry, "Groundwater," Prentice Hall, Englewood Cliffs, 1979.

27. S. Epstein and T. K. Mayeda, "Variations of ^{18}O Content of Waters from Natural Sources," Geochimica et Cosmochimica Acta, Vol. 4, No. 5, 1953, pp. 213-224.

28. M. L. Coleman, T. J. Shepherd, J. J. Durham, J. E. Rouse and G. R. Moore, "Reduction of Water with Zinc for Hydrogen Isotope Analysis," Analytical Chemistry, Vol. 54, No. 6, 1982, pp. 993-995.

29. T. B. Coplen, "Normalization of Oxygen and Hydrogen Isotope Data," Chemical Geology, Vol. 72, 1988, pp. 293-297.

30. C. Güler, G. D. Thyne, E. J. McCray and A. K. Turner, "Evaluation of Graphical and Multivariate Statistical Methods for Classification of Water Chemistry Data," Hydrogeology Journal, Vol. 10, No. 4, 2002, pp. 455-474.

31. C. E. Brown, "Applied Multivariate Statistics in Geohydrology and Related Sciences," Springer, Berlin, 1998, p. 248.

32. J. H. Ward, "Hierarchical Groupings to Optimize an Objective Function," Journal of American Statistical Association, Vol 58, No. 301, 1963, pp. 236-244.

33. B. A. Helena, M. Vega, E. Barrado, R. Pardo and L. Fernandez, "A Case of Hydrochemical Characterization of Alluvial Aquifer Influenced by Human Activities," Water Air and Soil Pollution, Vol. 112, No. 3-4, 1999, pp. 365-387.

34. R. P. Ashley and J. W. Lloyd, "An Example of the Use of Factor Analysis and Cluster Analysis in Groundwater Chemistry Interpretation," Journal of Hydrology, Vol. 39, No. 3-4, 1978, pp. 335-364.

35. S. X. Meng and J. B. Maynard, "Use of Statistical Analysis to Formulate Conceptual Models of Geochemical Behavior: Water Chemical Data from the Botucatu Aquifer in São Paulo State, Brazil," Journal of Hydrology, Vol. 250, No. 1-4, 2001, pp. 78-97.

36. R. E. Williams, "Statistical Identification of Hydraulic Connections between the Surface of a Mountain and Internal Mineralized Sources," Ground Water, Vol. 20, No. 4, 1982, pp. 466-478.

37. J.-H. Kim, R.-H. Kim, J. Lee, T.-J. Cheong, B.-W. Yum and H.-W. Chang, "Multivariate Ststistical Analysis to Identify the Major Factors Governing Groundwater Quality in the Coastal Area of Kimje, South Korea," Hydrological Sciences, Vol. 19, No. 6, 2005, pp. 1261- 1276.

38. C. Güler and G. D. Thyne "Hydrologic and Geologic Factors Controlling Surface and Groundwater Chemistry in Indian Wells-Owens Valley Area, Southern California, USA," Journal of Hydrology, Vol. 285, No. 1-4, 2004, pp. 177-198.

39. R. A. Johnson and D. W. Wichern, "Applied Multivariate Statistical Analysis," 3rd Edition, Prentice-Hall Inc., Englewood Cliffs, 1992.

40. O. Vaseli, A. Buccianti, C. De Siena, N. Coradossi and M. Angelone, "Geochemical Characterisation of Ophiolitic Soils in a Temperate Climate; a Multivariate Statistical Approach," Geoderma, Vol. 75, No. 1-2, 1997, pp. 117- 123.

41. S. Swanson, J. Bahr, M. Schwar and K. Potter, "Two- -Way Cluster Analysis of Geochemical Data to Constrain Spring Source Waters," Chemical Geology, Vol. 179, No. 1-4, 2001, pp. 73-91.

42. J. Taboada, A. Vaamonle, A. Saavedra and L. Alejano, "Application of Geostatistical Techniques to Exploitation Planning in Slate Quarries," Engineering Geology, Vol. 47, No. 3, 1997, pp. 269-277.

43. H. F. Kaiser, "An Index of Functional Simplicity," Psychometrika, Vol. 39, No. 1, 1974, pp. 31-36.

44. L. Plummer, J. Busby, R. Lee and B. Hanshaw, "Geochemical Modeling of the Madison Aquifer in Parts of Montana, Wyoming, and South Dakota," Water Resource Research, Vol. 26, No. 9, 1990, pp. 1981-2014.

45. W. M. Edmunds and P. Smedley, "Residence Time Indicators in Groundwater: The East Midlands Triassic Sandstone Aquifer," Applied Geochemistry, Vol. 15, No. 6, 2000, pp. 737-752.

46. S. Adams, G. Tredoux, C. Harris, R. Titus and K. Pietersen, "Hydrochemical Characteristics of Aquifers near Sutherland in the Western Karoo, South Africa," Journal of Hydrology, Vol. 241, No. 1-2, 2001, pp. 91-103.

47. L. Herczeg and T. E. Payne, "Recharge and Weathering Processes in Fractured Rock Aquifers in Northern Australia," In: Y. Kharaka and A. E. Maest, Eds., Water-Rock Interaction, Balkema, Rotterdam, 1992, pp. 561-564.

48. T. Coplen, A. L. Herczeg and C. Barnes, "Isotope Engineering-Using Stable Isotopes of the Water Molecule to Solve Practical Problems," In: P.G. Cook and A. L. Herczeg, Eds., Environmental

Tracers in Subsurface Hydrology, Kluwer Academic Pubs, Boston, 2000, pp. 79-110.

49. T. T. Akiti, "Geochemical and Isotopic Studies of Groundwater in Upper West Region of Ghana," Unpublished Ph.D. Thesis, Universitié de Paris Sud, Paris, 1980.

50. I. Chebotarev, "Metamorphism of Neutral Water in the Crust of Weathering," Geochemica et Cosmochimica Acta, Vol. 8, 1955, pp. 22-48, 137-170, 192-212.

7
CHAPTER

Mapping of Lineaments for Groundwater Targeting and Sustainable Water Resource Management in Hard Rock Hydrogeological Environment Using RS- GIS

Pothiraj Prabu[1] and Baskaran Rajagopalan[1]

[1] *Department of Industries and Earth Sciences, Tamil University, Thanjavur, Tamilnadu, India*

1. INTRODUCTION

Lineament definition and history

Numerous definitions of the term 'lineament' are given in the literature and various attributes are sometimes linked to the term - such as 'geologic lineament', 'tectonic lineament', 'photo lineament' or 'geophysical lineament' - either describing the assumed origin of the linear feature or sometimes the data source from which it has been derived. Some researchers also use the term 'fracture trace' or 'photo linear' as an alternative term. The work by Lattman and Parizek (1964) is commonly regarded as pioneering work in groundwater exploration; they mapped linear features (fracture traces) on stereo-pairs of aerial photographs in carbonate terrain in the eastern United States and subsequently showed the correlation between well productivity and distance to the identified features.

Lineament mapping was used long before this work in other geological applications and the first usage of the term lineament in geology is probably from a paper by Hobbs (1904, 1912), who defined lineaments as significant lines of landscape caused by joints and faults, revealing the architecture of the rock basement. This was later used by O' Leary et al. (1976) as a basis for developed definitions. Lineaments have been defined as extended mappable linear or curvilinear features of a surface whose parts align in straight or nearly straight relationships that may be the expression of folds, fractures or faults in the subsurface. These features are mappable at various scales, from local to continental, and can be utilized in mineral, oil and gas, and groundwater exploration studies.

Linear features on the Earth's surface have attracted the attention of geologists for many years. This interest has grown most rapidly in geological studies since the introduction of aerial photographs and satellite images. At the beginning, to the middle of the twentieth century, several geologists recognized the existence and significance of linear geomorphic features that were the surface expression of zones of weakness or structural displacement in the crust of the Earth.

Studies revealed a close relationship between lineaments and groundwater flow and yield (Mabee et al., 1994; Magowe and Carr, 1999; Fernandes and Rudolph, 2001). Generally lineaments are underlain by zones of localized weathering and increased permeability and porosity. Meanwhile, some researchers studied relationships between groundwater productivity and the number of lineaments within specifically designated areas or lineament density rather than the lineament itself (Hardcastle, 1995). Therefore, mapping of lineaments closely related to groundwater occurrence and yield is essential to groundwater surveys, development and management. In the last two decades remote sensing and GIS have been widely used for preparation of different types of thematic layers and their integration for different purposes.

This research work focuses on developing the remote sensing and Geographic Information Systems (GIS) methodology for regional groundwater potential evaluation. The objectives of this study are to (i) produce a regional structural lineament map of the study area from remotely sensed data, (ii) determine the hydro geological implication of the lineaments by integrating them with the available ancillary data (Digital Elevation Model [DEM] and geological map), (iii) analyse the lineament trend distribution of the study area using rose diagrams, lineament density maps and lineament intersection maps.

2. DESCRIPTION OF STUDY AREA

The Vaigai sub-basin extends over approximately 849 km^2 and lies between 09^0 30\` 00" and $10^0$ 00\` 00"N latitudes and 77^0 15\` 10"and $77^0$ 45 00\` E longitudes in the western part of Tamilnadu, India. It originates at the altitude of 1661m in the Western Ghats of Tamilnadu in the Theni district (Figure 1). The basin is generally hot and dry except during the winter season. The maximum and minimum temperature for the basin is 40.7 ^{0}C and 16.0 ^{0}C. The area receives an average annual rainfall of about 384 mm. The surface runoff goes to stream as instant flow. Rainfall is the direct recharge source and the irrigation return flow is the indirect source of groundwater in the Vaigai sub-basin. The study area depends mainly on the north-east monsoon rains which are brought by the troughs of low pressure established in the Bay of Bengal. Most of the farmers depend on the groundwater for their irrigational needs. There are a few tanks across these drainages, however, most of these remain dry.

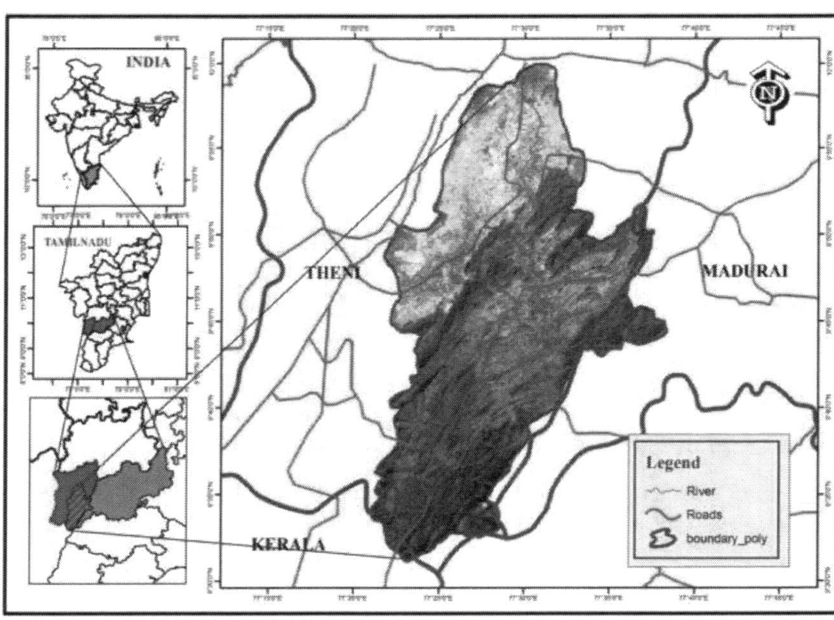

FIGURE 1. Location of the study area with ASTER 30m DEM

3. METHODOLOGY

The Indian Remote Sensing Satellite (IRS) ID, Linear Image Self-Scanning (LISS) III of geo-coded False Colour Composites (FCC), generated from the bands 2, 3 and 4 on 1:50,000 scale was used for the present study. The application of higher-resolution 30- m Advanced Space borne Thermal Emission and Reflection Radiometer (ASTER) imagery yielded better results in lineament interpretation compared to IRS 1D imagery due to improved spatial resolution. Lineament mapping is normally undertaken based on geomorphological features such as aligned ridges and valleys, displacement of ridge lines, scarp faces and river passages, straight drainage channel segments, pronounced breaks in crystalline rock masses and aligned surface depression For the study area, the main interest was topographically negative lineaments, which may represent joints, faults and probably shear zones (Juhari and Ibrahim 1997; Koch and Mather 1997; Solomon and Ghebreab 2006). To eliminate the non-geological elements, such as paths, roads, power cables and field boundaries in the study area, geographical maps and field checking were undertaken using the method suggested by Yassaghi (2006).

3.1. GEOLOGY

Eleven geologic features were identified and mapped by the Geological Survey of India, shown inFigure 2.

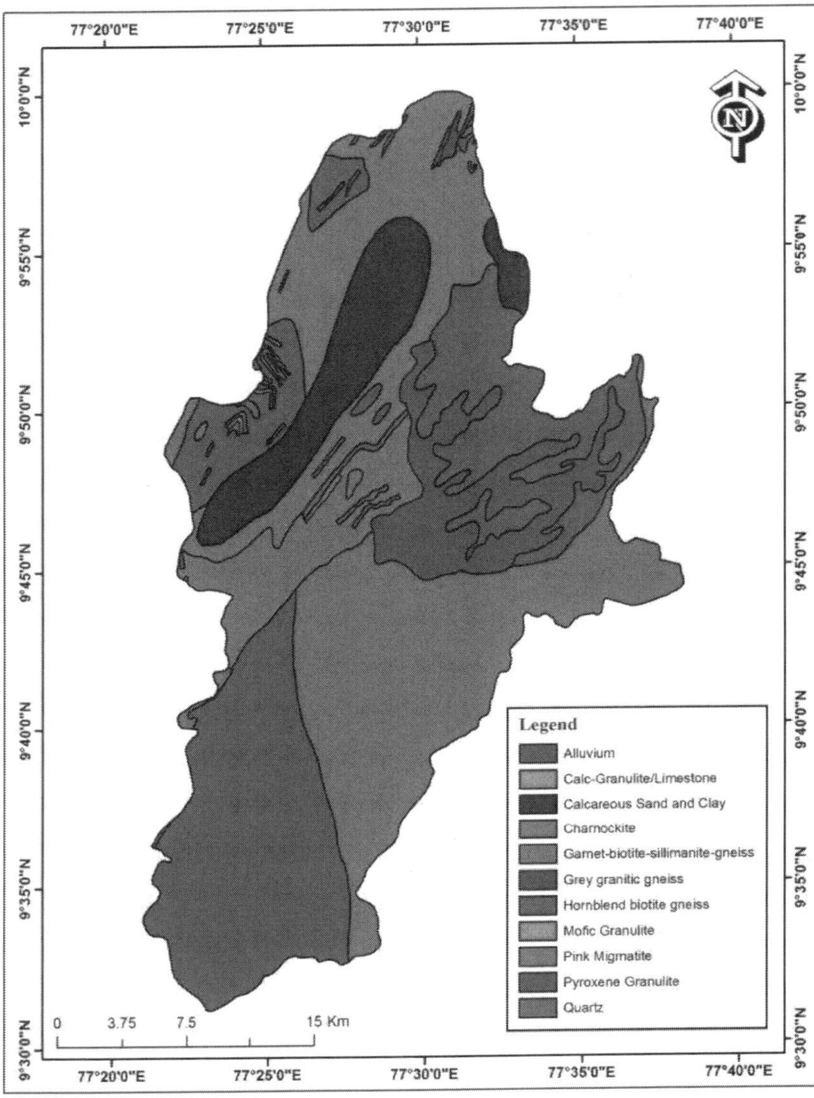

FIGURE 2. Geology of the study area

3.2. Lineament Analysis

The mapped structural lineaments were mapped and analysed using the lineament density (LD), lineament frequency (LF) and lineament intersection (LI) parameters. The results of the analysis are presented as the lineament map, lineament density map, rose diagram, lineament frequency map and lineament intersection map (Figures 3, 4, 5, 6 & 7) respectively.

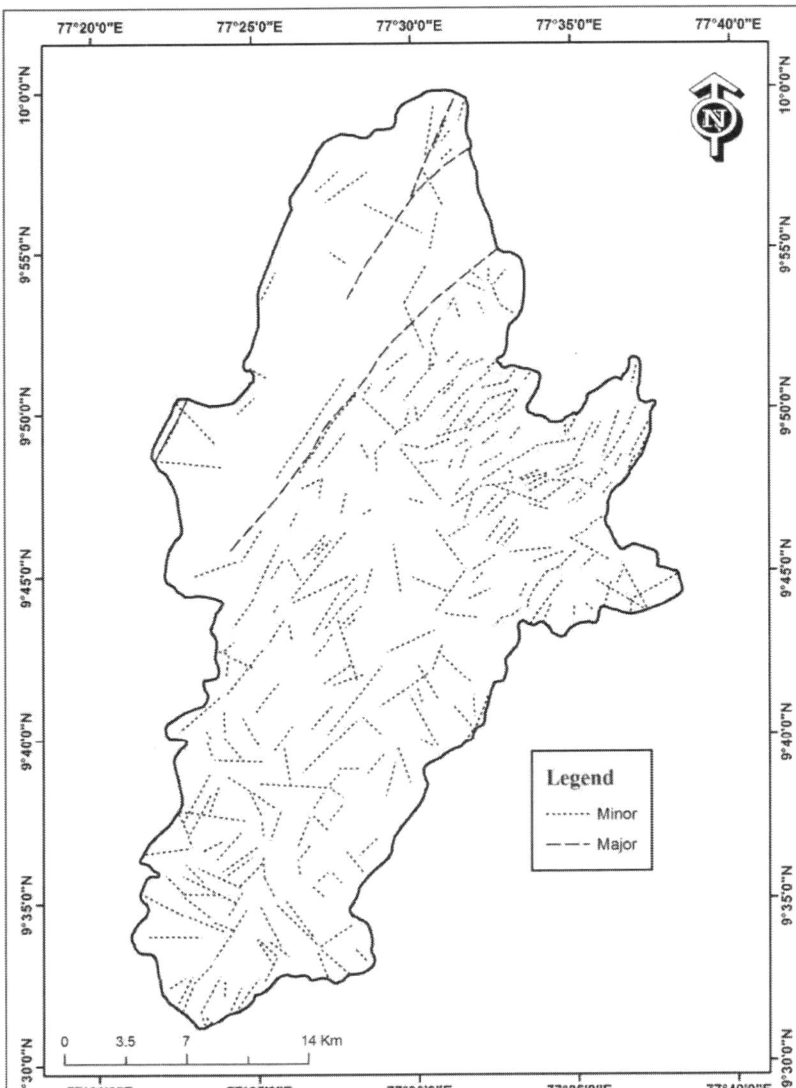

FIGURE 3. Lineament map of the study area

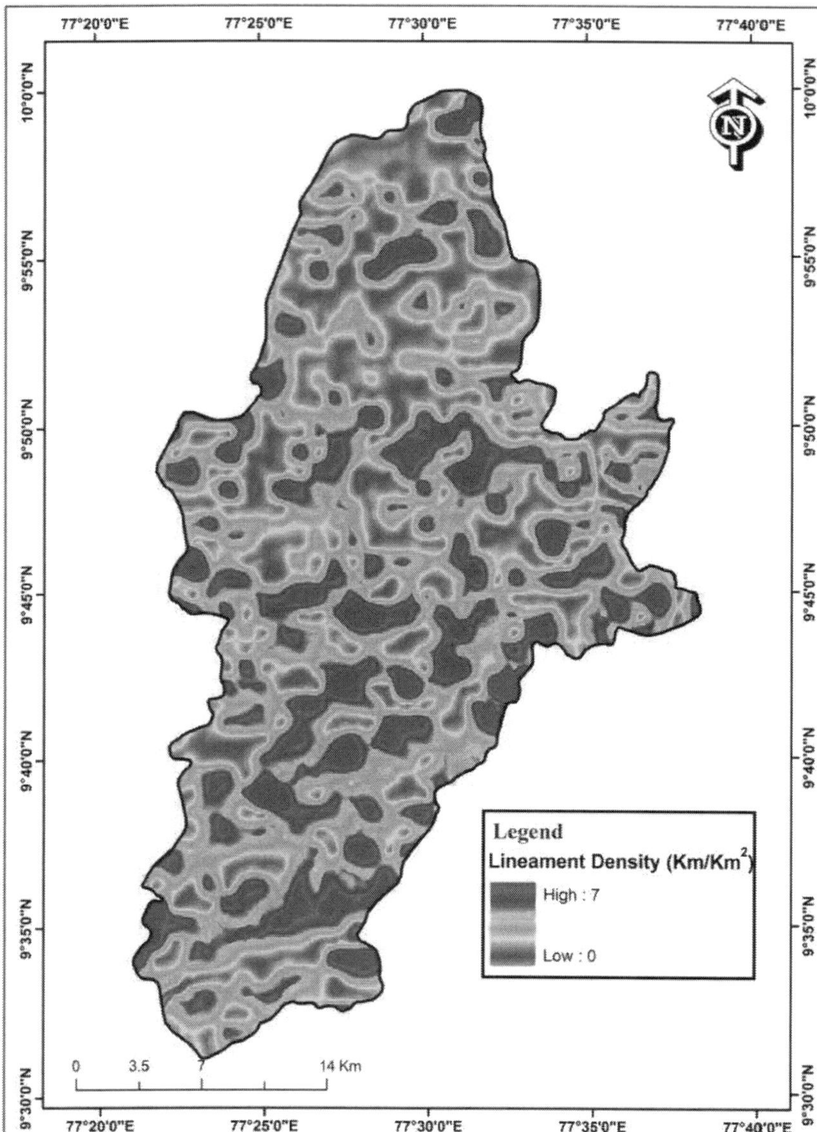

FIGURE 4. Lineament density of the study area

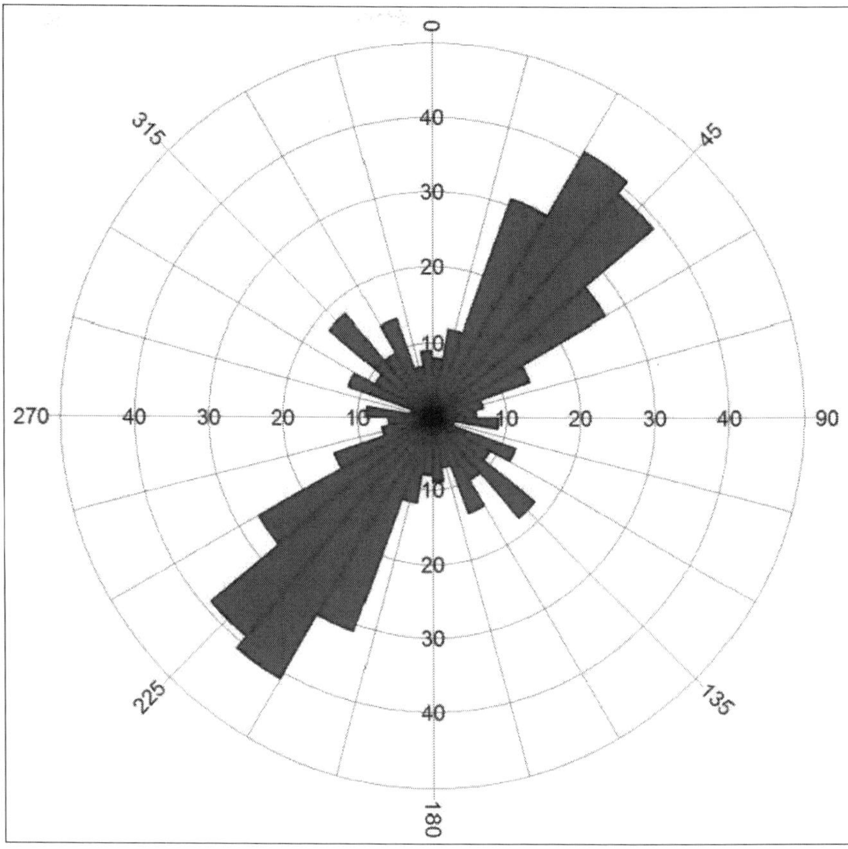

FIGURE 5. Lineament rose plot in the study area

4. RESULTS AND DISCUSSION

4.1. Frequency And Spatial Location Of The Fractures

The lineament and frequency map (Fig. 3 and Fig. 6) shows that the lineaments/fracture distribution is hardly homogeneous. The lineament density variation map (Fig. 6) shows the lineament numbers to be in the range of 0 and 7. The majority of the fractures are located on lithologies that correspond to the term "hard rocks", which generally refers to igneous and metamorphic rocks (Krasny 1996, 2002). Therefore, the discussed character represents an initial indication for the unified tectonic and hydrogeologic behaviour of the hard rock environment. The majority of the lineaments/fractures are located on the Hornblend biotite gneiss and minority of the lineaments/fractures are located on the Pink Migmatite (Table 1).

4.2. Orientation Of The Lineaments

The orientation of the lineaments is analysed by constructing rose diagrams (Fig.5). Even though these diagrams are not length-weighted, they can indicate on each occasion what the most dominant directions of the fractures are. This analysis is very critical for the study of groundwater flow, as in most cases the orientation of the fractures is identical to the orientation of the preferential flow path.

The faults rose plot indicates two sets of orientation classes. The main two classes have NE and SW strike, while others have NW and SE strike. The uniformity of fracture orientation becomes an additional indication for the hydrogeologic regime.

4.3. Size Of The Lineaments

Fracture dimensions (aperture and apparent aperture) are very difficult to define and the depth of the apertures makes the measurements even more complicated. Nevertheless, length measurements can be taken relatively easily and they are also significant, since a fracture with a greater length affects the groundwater flow in a more dominant way than those of smaller length. The calculated total length of lineament/fracture per unit area in each lithology are shown in Table 1.

TABLE 1. Total length of lineaments in each geological feature

No	Description	Frequency	Percent (%)	Length (Km)	Percent (%)
1	Hornblend biotite gneiss	163	44.05	221.68	45.51
2	Charnockite	104	28.11	158.27	32.49
3	Alluvium	54	14.59	57.81	11.87
4	Garnet-biotite-sillimanite-gneiss	26	7.03	28.60	5.87
5	Calcareous Sand and Clay	7	1.89	14.17	2.91
6	Quartz	11	2.97	3.17	0.65
7	Pyroxene granulite	1	0.27	1.43	0.29
8	Grey granitic gneiss	2	0.54	1.16	0.24
9	Calc-granulite/Limestone	1	0.27	0.55	0.11
10	Pink migmatite	1	0.27	0.28	0.06
	Total	370	100.00	487.13	100.00

4.4. Density Of The Lineaments

The purpose of the fracture density analysis is to calculate frequency of the fractures per unit area. With this analysis a map has been produced showing concentrations of the lineaments over the study area (Fig. 4). The map in Figure 3 shows that very high density is observed in areas of Hornblende biotite gneiss and Charnockite (7 Km/ Sq.km^2), indicates the high degree of hydraulic interconnection between the above lithologic units as surface water circulates through these discontinuities. This is verified in the next consideration (degree of fractures intersection). On the other hand, very low density is observed in calcareous sand and clay (1 Km/ Sq.km^2), and quartz in areas where combination of more lithological features dominate. This verifies that these lithologies are affected by tectonic activity.

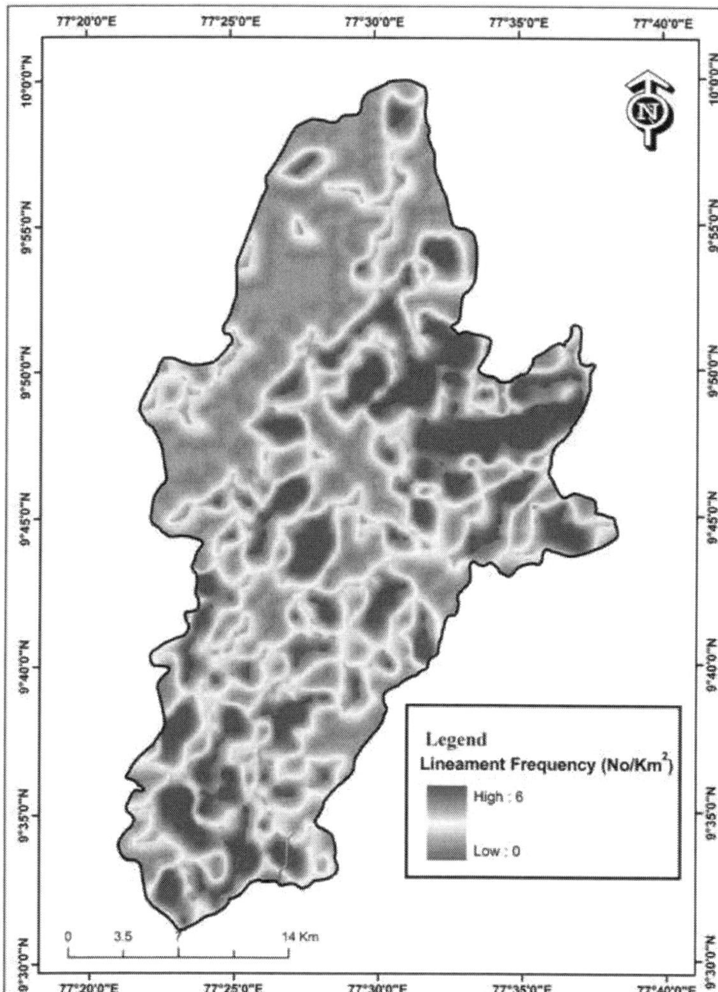

FIGURE 6. Lineament frequency of the study area

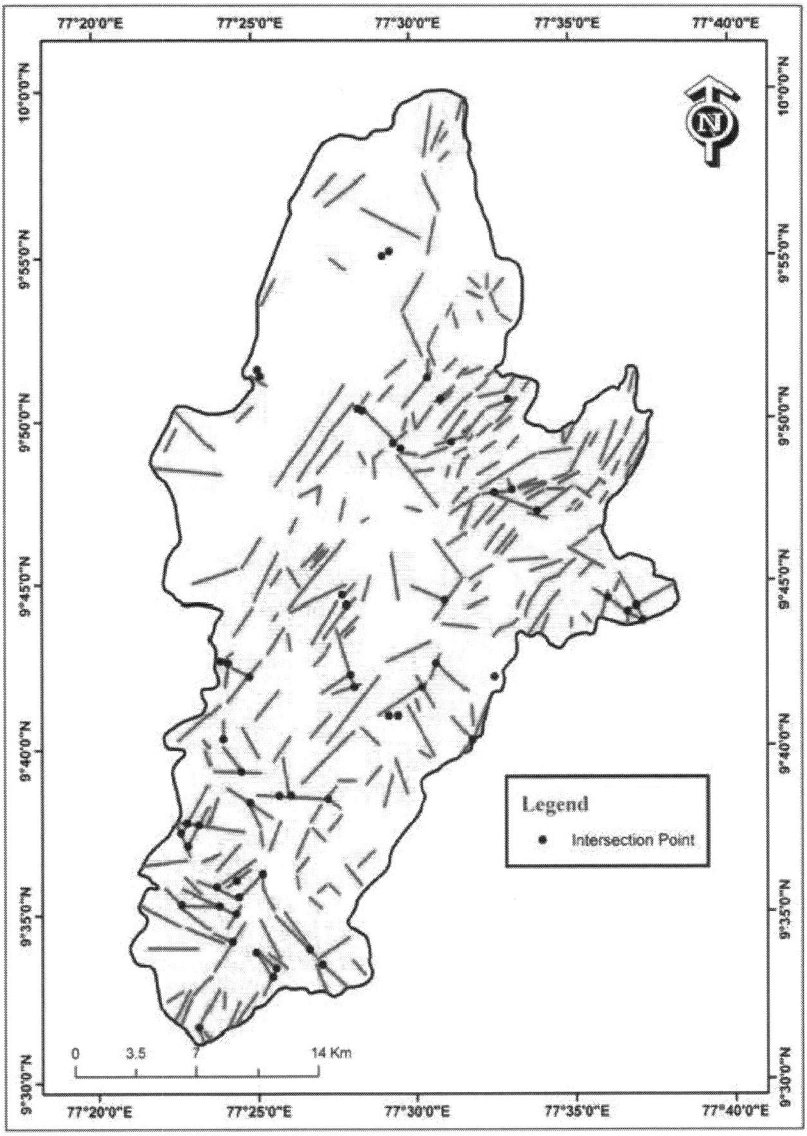

FIGURE 7. Lineament intersection of the study area

4.5. Degree Of Lineaments Intersection

The density of lineaments along with the degree of lineament intersection determine the degree of anisotropy of groundwater flow in the fracture network, as in environments with a high degree of interconnection where groundwater flow is smoother and more uniform. Fracture intersection density is a map showing the frequency of intersections that occur in a unit area. The purpose of using intersection density maps is to estimate the areas of diverse fracture orientations. If the fractures do not intersect in an area, the resultant map will be represented by a plain map with almost no density contours and the fractures are almost parallel or sub-parallel in an area. The lineament intersection map of the study area (Fig. 7) indicates high and very high intersection in the same areas where there is very high density of lineaments. The zones of high lineament intersection over the study area are feasible zones for groundwater potential evaluation.

5. CONCLUSIONS

Groundwater studies on hard formations often require extraction of data from images and remote sensing, and GIS. Due to insufficient data, maps of lineament and structural elements are important tools that may reveal points of groundwater recharge and discharge, flow and development. In particular, groundwater occurrences in hard formations are mainly controlled by the lineaments corresponding to fractures, joints and faults. Furthermore, the distribution of lineament is closely related to groundwater discharge points and their concentration.

Remote sensing has proved to be a useful tool in lineament identification and mapping. This study demonstrates the application of remotely sensed data for lineament interpretation in a hard rock hydrogeological environment. A Digital Elevation Model (DEM) was generated to improve the interpretation. The lineament analysis has been effectively done in a GIS environment. Thematic maps, such as lineament frequency, lineament density and lineament intersection, were prepared using the interpolation technique.

The results from the study show that the remote sensing technique is capable of extracting lineament trends in an inaccessible tropical forest. The study has led to the delineation of areas where groundwater occurrences are most promising for sustainable supply, suggesting where further geophysical surveys can be concentrated. It is therefore suggested that the high lineament intersection and density should be combed with detailed geoelectrical surveys for quantitative evaluation of the groundwater potential of the study area. Properly sited wells in drought-stricken areas could change the lives of many and the remote-sensing analysts and lineament interpreters around the world are without doubt important in this process.

REFERENCES

1. A. J Fernandes, D. L Rudolph, 2001The influence of Cenozoic tectonics on the Groundwater production capacity of fractured zones: A case study in Sao Paulo, Brazil. Hydrogeology Journal 9 151 167

2. K. C Hardcastle, 1995Photolineament factor: A new computer-aided method for remotely sensed fractured. Photogrammetric Engineering & Remote Sensing 61 (6), 739-747.

3. W. H Hobbs, 1904Lineaments of the Atlantic border region. Geological Society of America Bulletin 15 483 506

4. W. H Hobbs, 1912Earth Features and Their Meaning: An Introduction to Geology for the Student and General Reader. Macmillan Co., New York, 347.

5. M. A Juhari, and A Ibrahim, 1997Geological Applications of Landsat Thematic Mapper Imagery: Mapping and Analysis of Lineaments in NW Peninsula Malaysia. ACRS. Available online at: www.gisdevelopment.

6. M Koch, and P. M Mathar, 1997Lineament mapping for groundwater resource assessment: a comparison of digital Synthetic Aperture Radar (SAR) imagery and stereoscopic Large Format Camera (LFC) photographs in the Red Sea Hills, Sudan. International Journal of Remote Sensing, 27 4471 4493

7. J Krasny, 2002Hard Rock Hydrogeology. 1st Workshop on Fissured Rocks Hydrogeology Proceedings, Athens, 11 18

8. J Krasny, 1996Hydrogeological Environment in Hard Rocks: An attempt at its schematizing and terminological consideration. Acta Univesitatis Carolinae Geologica, 40 115 122

9. L. H Lattman, R. R Parizek, 1964Relationship between fracture traces and the occurrence of groundwater in carbonate rocks. Journal Hydrology 2 73 91

10. S. B Mabee, K. C Hardcastle, D. U Wise, 1994A method of collecting and analyzing lineaments for regional-scale fractured-bedrock aquifer studies. Ground Water 32 (6), 884-894.

11. M Magowe, J. R Carr, 1999Relationship between lineaments and ground water occurrence in western Botswana. Ground Water 37 (2), 282-286.

12. O Leary, D. W Freidman, J. D Pohn, H. A Lineaments, linear, lineation-some proposed new standards for old terms. Geological Society of America Bulletin 87 1463 1469

13. S Solomon, and W Ghebreab, 2006Lineament characterization and their tectonic significance using Landsat TM data and field studies in the central highlands of Eritrea. Journal of African Earth Sciences, 46 371 378

14. A Yassaghi, 2006Integration of Landsat imagery interpretation and geomagnetic data on verification of deep-seated transverse fault lineaments in SE Zagrosa, Iran International Journal of Remote Sensing, 27 4529 4544

8

CHAPTER

Geological Control upon Groundwater Flow and Major Ion Chemistry with Influence on Basin Management in a Coastal Aquifer, South Australia

Nara Somaratne, Jacqueline Frizenschaf

South Australian Water Corporation, 250 Victoria Square, Adelaide, Australia

ABSTRACT

Estimation of natural recharge and potential for seawater intrusion are critical considerations for management of coastal freshwater aquifers. We show hydrochemical signatures of groundwater to identify the influence of geological control on chemical processes in a coastal groundwater system. We used dominant hydrochemical facies, salinity and magnesium ions to determine two main groundwater flow paths with different origins and ages. Mixing of groundwater with different origins and ages results in unreliable recharge estimates using chlorofluorocarbon (CFC) and chloride mass balance (CMB) methods, thus limiting available methods for recharge assessment.

Interpretation of hydrochemical data suggests that calcium carbonate dissolution, ion exchange processes and mixing with sea aerosol in coastal zones are the main influencing factors on groundwater chemistry. Restricted groundwater flows due to occurrence of a basement high at the southern side of the basin boundary influence the distance to the toe of the saline wedge. Thus, knowledge of geological control over groundwater systems forms an important part of characterising basins and contributes toward effective management of groundwater resources.

Keywords: Hydrochemistry; Groundwater Mixing; Groundwater Management; Recharge

1. INTRODUCTION

Spatial distribution of groundwater chemistry is primarily influenced by mineralogical composition of sediments in contact with groundwater [1]. The nature and distribution of aquifers and aquitards in a geologic system are controlled by the lithology, stratigraphy and structure of geologic formations. The main geological controls are the stratigraphy and structural features such as cleavages, fractures, faults, and solution features which are the geometrical properties of the geologic systems. The influence of geological control changes natural water chemistry and therefore knowledge of main hydrochemical processes that control groundwater chemistry is a prerequisite for effective basin management [2]. Several studies examined the influence of geological controls: the presence of faults [3,4]; geology and structure of bedrock [5,6]; fracture control [7] which influences groundwater flow and water chemistry; and hydrogeological study of groundwater flow diversion caused by anticlinal structures [8].

Among geological controls, the presence of faults and their influence on flow paths and water chemistry are widely reported and these are known to be effective flow barriers [3]. Green et al. [4] reported flows across regional scale faults and found that groundwater flow either enhanced or inhibited, depended on orientation of fractures with respect to the cross fault hydraulic gradient. Bensel et al. [9] and Meeder [10] related the impact of faults on near surface processes to surface features such as vegetation patterns and drainage networks; and thus, faults can have great influence on transport processes in the sub-surface and are important in water management.

Eaton et al. [7] noted that there are few studies on the hydrogeology of sedimentary rock aquitards, although they are important controls in regional ground water flow systems. Cherry et al. [11] highlighted that determination of aquitard integrity is an important aspect of groundwater management because it influences water quality. Thus, the presence or absence of an aquitard, which is not necessarily a confining unit, can have a significant impact on basin water quality. We examined effects of basement rock outcrop and discontinuity of the clay aquitard on groundwater salinity, water chemistry, influence on recharge estimation methods and extent of the saline wedge in a coastal aquifer in South Australia.

2. STUDY AREA

The Uley South basin is located in the Southern Eyre Peninsula in South Australia (**Figure 1**). Average annual rainfall and pan evaporation is 550 mm and 1550 mm, respectively. The basin is used for town water supply with current allocation of about 6.8×10^6 m^3 per year. The geology and hydrogeology of the basin is characterised based on: Segnit [12], Johns [13], Morton and Steel [14], Wilson [15], and Shepherd [16]. These works include development of a geological model from borehole information, geological surface mapping, delineation of saturated limestone boundaries and pumping tests with attention focussed on Uley Wanilla, Uley East and Uley South groundwater lenses.

The aquifer system in Uley South is comprised of Quaternary Bridgewater Formation limestone (QL aquifer), beneath this is the Upper Tertiary Clay (Uley Formation) unit with thickness of 5 - 25 m, which forms an aquitard between the Tertiary Sand (TS) of Wanilla Formation and the Quaternary aquifer systems [14]. Recent studies [17,18] show that there is a hydraulic connection between the Uley South unconfined QL aquifer and the Uley Wanilla and Big Swamp areas through the TS aquifer (Figures 1 and 2). The basin is topographically closed and the low-lying central portion contains numerous sinkholes. Runoff is highly ephemeral, occurs only after moderate to high intensity rainfall and persists only tens to hundreds of meters before entering a sinkhole [17, 19,20]. Surface drainage is poorly developed as is typical of semi-arid karstic terrains that support only thin soils combined to produce scant vegetation. The groundwater flow direction in Uley South is from north-east to southwest.

An important surface feature is the Big Swamp (Figure 1) with an underlying Tertiary Clay with the primary mechanism of discharge through evaporation. In wetter years, Big Swamp overflows and infiltrates into the Quaternary Limestone unconfined aquifer and leaks into the TS aquifer raising salinity up to 5200 mg/L near Big Swamp.

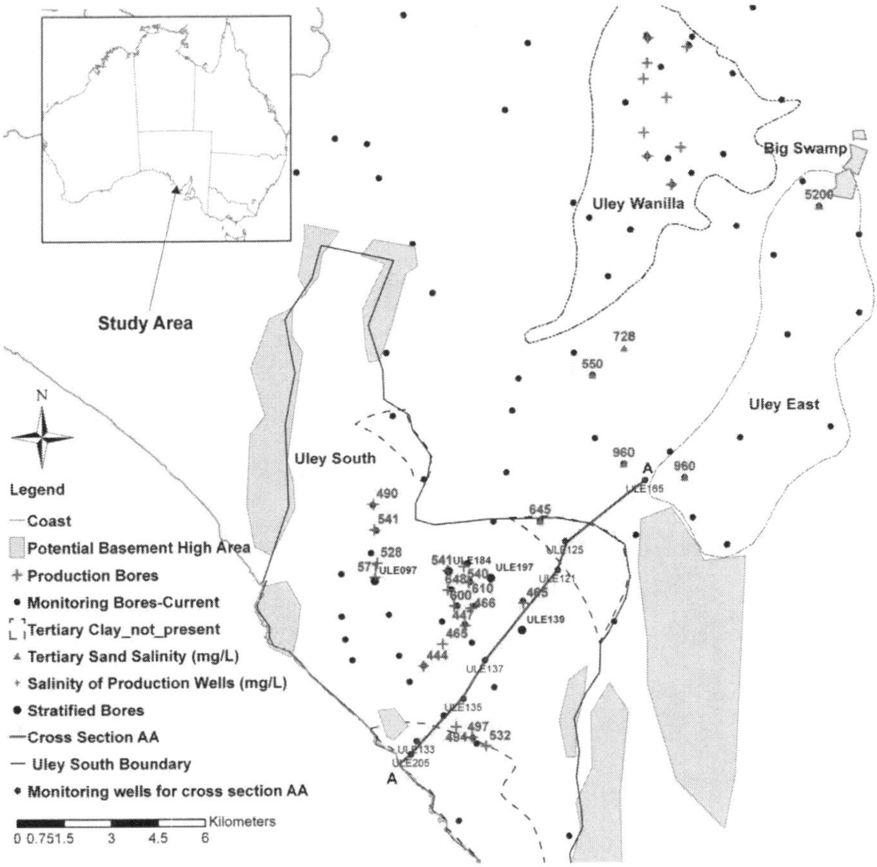

Figure 1. Location of the study area.

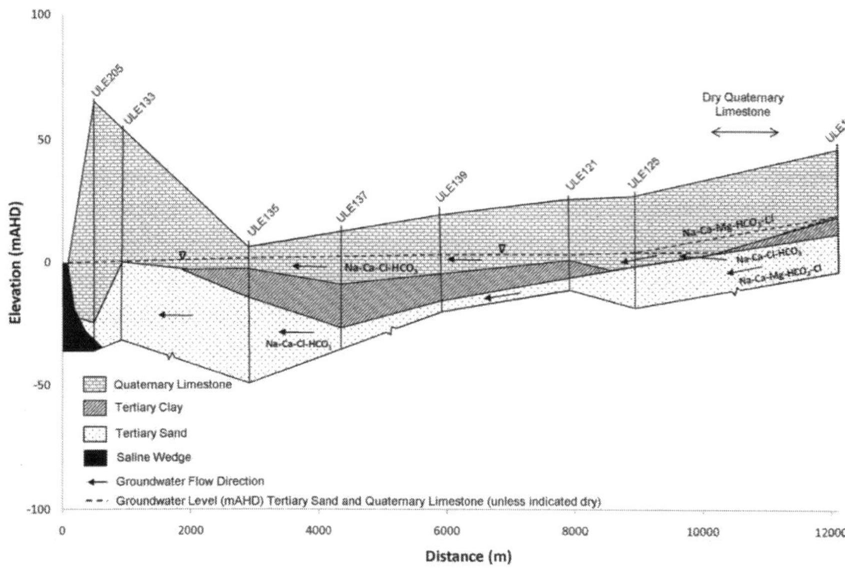

Figure 2. Cross section AA from the coast in the Uley South basin to Uley East basin's western edge.

3. RESULTS AND DISCUSSION

The influence of geological control is divided into three sections: influence on QL aquifer salinity, hydrochemistry of QL groundwater, and the effect of extension of the saline wedge in TS aquifer. These are important factors in managing the Uley South basin.

3.1. Influence on Salinity

In Uley South, the occurrence of basement high at the basin boundary and discontinuity of aquitards and solution features constitute the most important controls upon salinity and water chemistry. Salinity measurements taken from the town water supply (TWS) wells indicate higher salinity in the north-central portion than elsewhere in the basin (**Figure 1**). Salinity distribution in the basin is influenced by two main geological controls: the occurrence of basement high in the southern half of the basin, and the area of Tertiary Clay absence in the central part (Figures 1 and 2).

Harrington et al. [17] and Somaratne [18] show that the high-salinity plume of groundwater in the TS aquifer originates from the Big Swamp area as a result of downward leakage of Big Swamp's surface water through the clay aquitard. As the plume moves down gradient along its main flow path, the western edge of the Uley East basin, it is diluted from 5200 mg/L salinity at Big Swamp to 960 mg/L at the southern boundary of the Uley East basin. The area where aquitard is absent in Uley South, the TS aquifer receives direct recharge, resulting in further dilution of salinity to 645 mg/L (**Figure 1**). As a result, pronounced salinity stratification is found in QL aquifer monitoring well ULE197, down-gradient to the Tertiary Clay absence area (**Figure 3**(a)). In the ULE197 well, salinity of the QL base is identical to that in TS aquifer monitoring well ULE109, located at the same site. Harrington et al. [17] indicated that residence time in Big Swamp and Uley Wanilla groundwater age is >45 years. Mixing of old waters originating from Uley Wanilla and Big Swamp with younger water (<20 years) in Uley South has important ramifications on applicable recharge estimation methods. When mixing occurs of different groundwater systems with different ages and different chlorides concentrations, this results in unreliable recharge estimates using chlorofluorocarbon [21,22] and chloride mass balance methods [23,24], thus limiting the methods available for recharge assessment.

In contrast, the southern half of the QL aquifer has no link to the Big Swamp area through either QL or TS aquifers due to occurrence of the basement high at the Uley South boundary (**Figure 1**). Therefore, QL water in the southern part of Uley South is primarily derived from direct rainfall recharge within Uley South. ULE139 is a QL aquifer monitoring well, which shows no evidence of salinity stratification (**Figure 3**(b)).

3.2. Influence on Hydrochemical Signatures

The combination of aquifer heterogeneity, low hydraulic gradient, recharge through solution features in addition to diffuse recharge, mixing with different water types and modification along flow paths results in Uley South being hydrochemically complex. Hydrochemistry was assessed principally by major ion chemistry within the Uley Wanilla, Uley East and Uley South lenses, and how these are related to evolution of hydrochemical processes in the Uley South basin.

Groundwater can be classified into "water types" on the basis of the absolute and relative abundance of major ions [25]. However, due to the predominance of Na and Cl ions in both QL and TS aquifers, the major ion differences present are inconclusive in distinguishing between QL and TS waters. For all QL and TS groundwaters, Ordens et al. [20] has shown that Cl/Br in both rainfall and groundwater samples are close to the Cl/Br ratio of seawater indicating groundwater chloride are marine derived implying that both QL and TS waters may have similar chemical histories.

Discrimination based on relative major ion abundance is more successful where recharging freshwater is of continental (Ca-

HCO_3 type) rather than marine (Na-Cl type) origin as in Uley South. QL and TS groundwater are analysed for pH, electrical conductivity, anions (HCO_3^-, SO_4^{2-}, Cl^-) and cations (Ca^{2+}, Mg^{2+}, Na^+, K^+). Although the majority of water has no dominant ion composition, Uley Wanilla water contains distinctively high Mg ions (**Figure 4**) as observed by Harrington et al. [17]. High chloride in TS waters is derived from the Big Swamp area and the Na-Cl type water is diluted towards Na-Cl-HCO_3 waters at Uley South. This represents either waters that have chemically evolved along flow paths or that result from mixing between Na-HCO_3 waters and Na-Cl waters.

Classification of groundwater into water types on the basis of major ion composition and other selected characteristics provides a framework for interpretation of groundwater pathways [26]. Eighty-five groundwater samples were used for QL aquifer. This includes the data of Evans [19] and town water supply wells' water quality data. For the TS aquifer, twenty-seven water samples were taken from Evans [19].

Hydrochemical examination shows that carbonate dis solution and ion exchange are the two dominant chemical processes active in the QL aquifer. Water types are categorised into four types of groundwater, which represents a chemical evolution from Na-Mg-Cl-HCO_3 groundwater at Uley Wanilla lens to end member Na-Cl groundwater at Big Swamp (**Figure 5**).

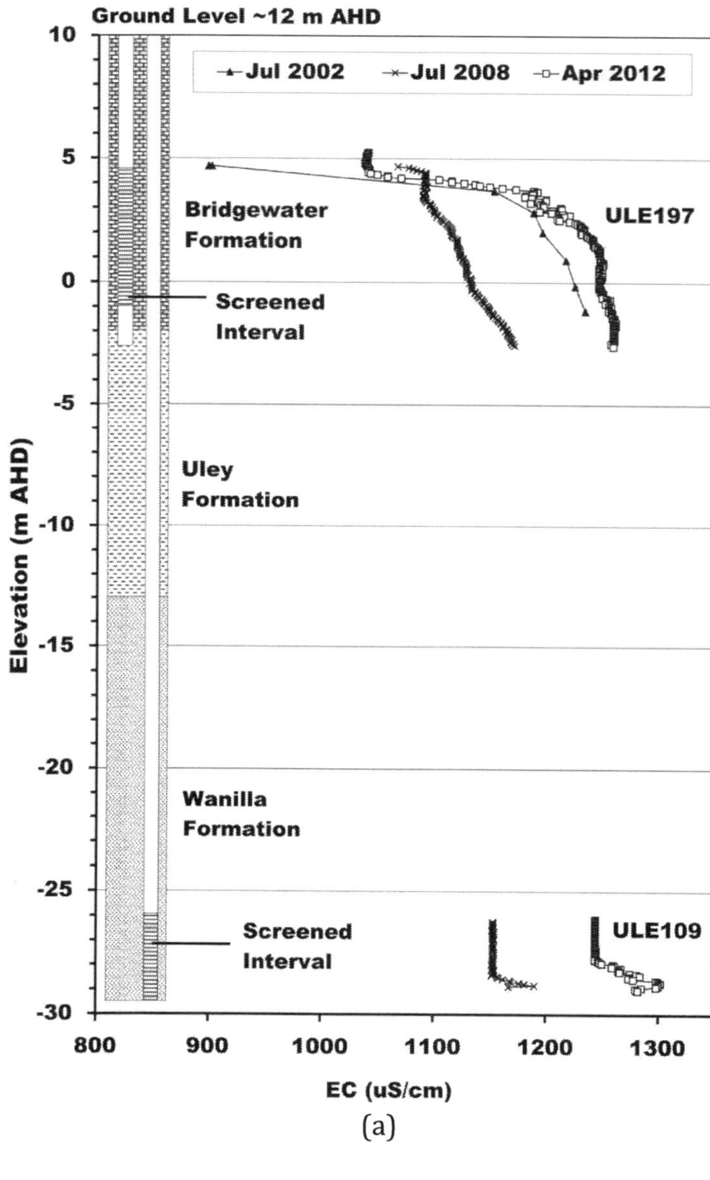
Ground Level ~12 m AHD
Jul 2002
Jul 2008
Apr 2012
Bridgewater Formation
ULE197
Screened Interval
Uley Formation
Wanilla Formation
Screened Interval
ULE109
Elevation (m AHD)
EC (uS/cm)
10
5
0
-5
-10
-15
-20
-25
-30
800
900
1000
1100
1200
1300

(a)

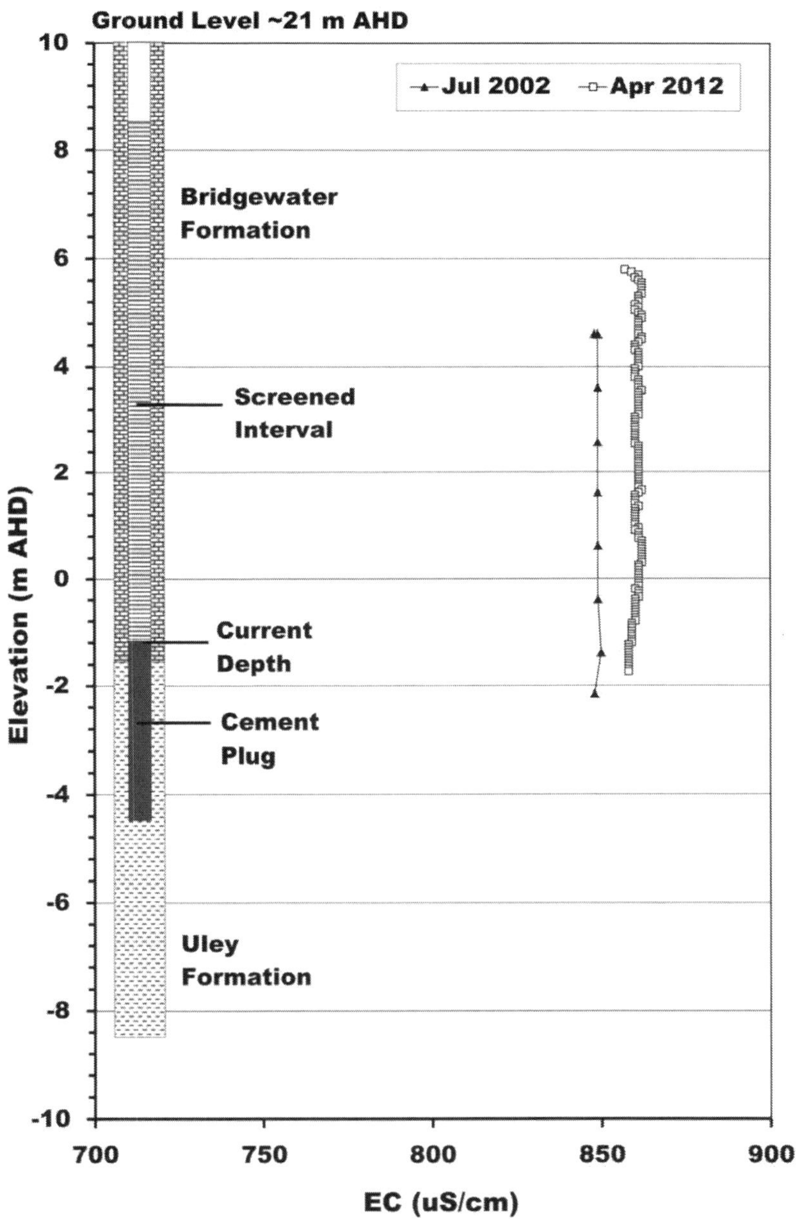

(b)

Figure 3. Salinity stratification of QL groundwater in the central part of the basin. (a) ULE109 and ULE197; (b) ULE139.

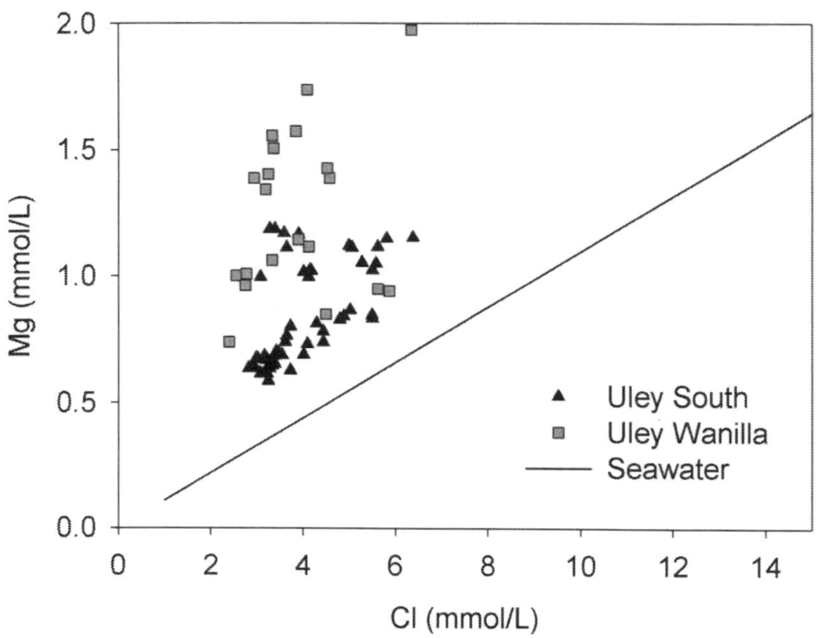

Figure 4. Cl vs Mg in Uley Wanilla and Uley South lenses.

3.2.1. Na-Cl or Na-Ca-Cl (1)

In the absence of evaporites, this type of water occurs in the Big Swamp due to enrichment of Na and Cl ions though evaporation. The end member of this water type is Na-Ca-Cl-HCO_3 or Na-Ca-HCO_3-Cl in the north-central part of Uley South basin, where salinity of town water supply wells are about 650 mg/L; about 100 mg/L higher than elsewhere in the QL aquifer.

3.2.2. Na-Mg-Ca-Cl-HCO_3/Na-Ca-Mg-HCO_3-Cl (2)

Na-Mg-Ca or Na-Ca-Mg water indicates that dissolution of dolomite (or dolomitic limestone) is taking place in the catchment area. In

Quaternary Limestone, this type of water occurs in the Uley Wanilla basin (**Figure 6**). In the TS aquifer, this type of water occurs either within the Uley Wanila basin or directly down-gradient in the flow direction, indicating direct leakage (interconnection between two aquifers) from QL to TS aquifer. This type of water in the Uley South QL aquifer is found only in production bores directly down gradient to Tertiary Clay absence area as a result of QL aquifer receiving TS waters. Apart from production wells directly down gradient, none of the monitoring wells nor any other town water supply wells contain this type of water.

3.2.3. Ca-Na-HCO_3-Cl/Ca-Na-Mg-HCO_3-Cl (3)

This is the typical water type found in limestone aquifers. The Ca-Na-HCO_3-Cl water type represents recently recharged water with an abundance of dissolved CO_2; thus calcite dissolution is taking place. This water type occurs in the southern part of the central basin where there is no TS water mixing with Uley South QL aquifer water. Uley South basin being a karstic limestone aquifer, sink holes collect runoff containing organics; this provides sufficient CO_2 for the following reactions to dissolve calcite.

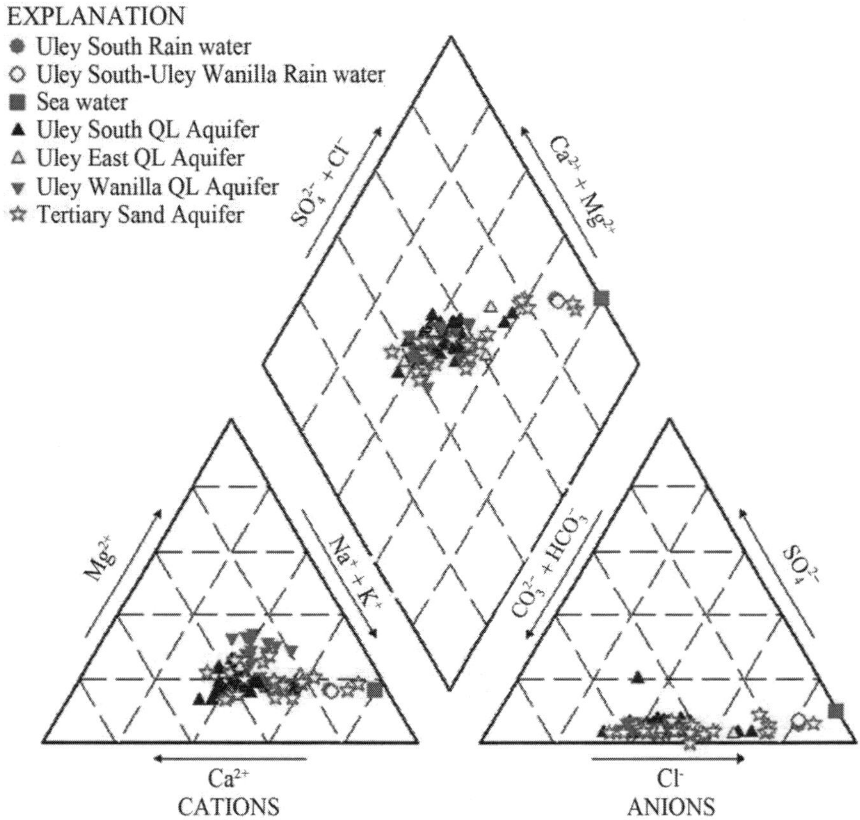

Figure 5. Piper diagram showing water types.

The main mechanism for dissolution of calcite is as follows:

$$CaCO_3 + CO_2 + H_2O = Ca^{2+} + 2HCO_3^- \quad (1)$$

The above reaction includes the following steps:

$$CO_2 + H_2O = H^+ + HCO_3^- \quad (2)$$

$$CaCO_3 + H^+ = Ca^{2+} + HCO_3^- \quad (3)$$

3.2.4. Na-Ca-Cl-HCO_3/Na-Ca-HCO_3-Cl (4)

Na-Ca-HCO_3-Cl water types represent Ca-HCO_3 water affected by mixing with Na-Cl waters or that has undergone cation exchange processes. The most common occurrence of such water is in the central basin (**Figure 6**), where high salinity waters originating from Big Swamp enter the QL aquifer. In addition another possible reason for the occurrence of this water type is cation exchanges, along with sea aerosol in the coastal zone.

3.3. Influence on Coastal Salinity

Seawater intrusion is generally identified through chemical analysis of groundwater. No single analysis definitively identifies seawater intrusion [27]; however by examining various analysis methods, it is evident when fresh groundwater mixes with seawater. According to HydroMetric WRI [27], at low chloride concentrations, it is often difficult to identify incipient seawater intrusion. This is due to the natural variation in fresh water chemistry at chloride concentrations below 1000 mg/L.

Mixing trends between groundwater and seawater are more easily defined when chloride concentrations exceed 1000 mg/L, as is evident in **Figure 7**where mixing increases with depth. As shown in **Figure 7**, groundwater mixing takes place in coastal aquifer monitoring wells from Na-Ca-Cl-HCO_3 to Na-Cl water types at two different distances from the coast at the same elevation. On the northern side, TS groundwater is derived from downward leakage from the Uley

Wanilla-Big Swamp system, which is approximately 22 km from the coastline. TS receives further direct recharge from areas where Tertiary Clay is absent.

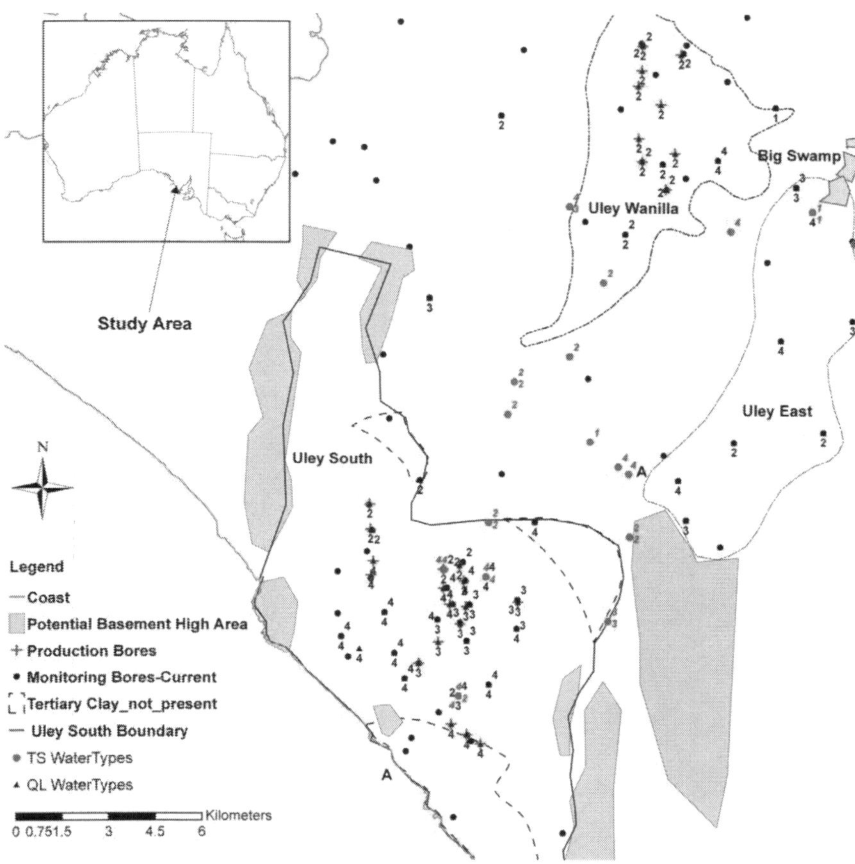

Figure 6. QL and TS aquifer water type distribution (1 = Na-Cl or Na-Ca-Cl; 2 = Na-Mg-Ca-Cl-HCO_3 or a-Ca-Mg-HCO_3-Cl or Ca-Na-Mg-HCO_3-Cl; 3 = Ca-Na-HCO_3-Cl or Ca-Na-Cl-HCO_3; 4 = Na-Ca-Cl-HCO_3 or Na-Ca-HCO_3-Cl).

Therefore, it is conceptualised that a large quantity of TS groundwater flows on the northern side with a higher hydraulic gradient (0.00458). This makes the distance to the toe of the saline wedge in TS smaller, less than 500 m on the northern side. In contrast, the TS aquifer on the southern side receives leakage from the QL aquifer within Uley South basin. The extent of the TS is limited only to the Uley South boundary, which is on average approximately 5 km from the coast, yielding an average hydraulic gradient of 0.00117. This results in comparatively small groundwater flows through TS along the southern side, creating a longer saline wedge for pressure balance at the same elevation. Distance to the toe of the saline wedge is about 900 m from the coastline on the southern side. In Uley South coastal aquifer monitoring wells, sea water is intercepted in Tertiary Sand aquifer in SLE069 at −23 m AHD and in ULE205 at −30 m AHD (**Figure 7**(b)) showing increasing salinity with depth. No QL aquifer coastal bores intercept saline waters indicating that the saline wedge in QL aquifer is less than 500 m.

4. CONCLUSIONS

We highlight the importance of integrating knowledge of hydrochemistry of groundwater sources into estimation of recharge and the potential for saline intrusion, both of which are essential components in assisting with effective management of groundwater resources. In particular, we analysed the groundwater chemistry from the Uley South coastal aquifer in South Australia, and examined the processes that control groundwater chemistry. The major outcome of this study is identification of two distinct groundwater sources due to geological controls and

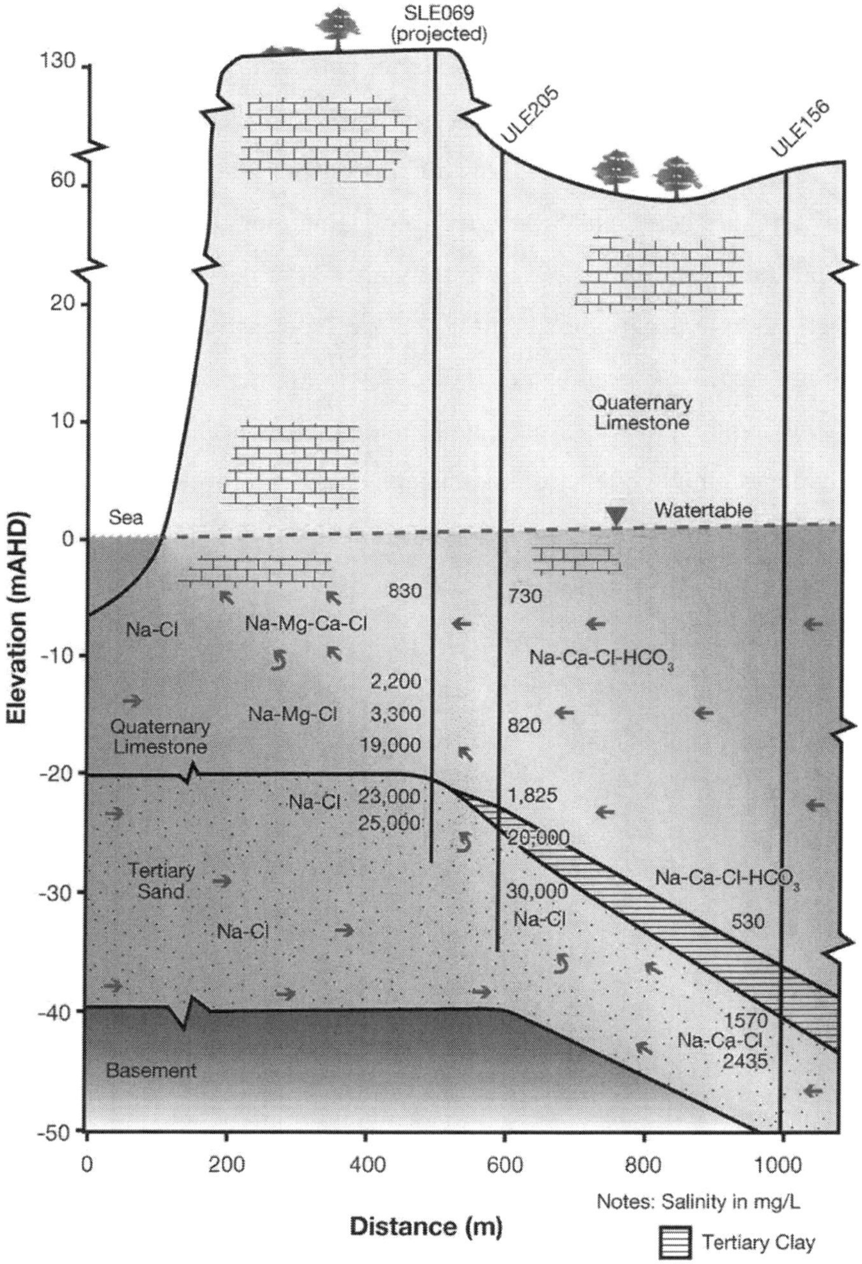
SLE069
(projected)
ULE205
ULE156
130
60
20
10
0
-10
-20
-30
-40
-50
Elevation (mAHD)
Sea
Watertable
Quaternary
Limestone
Na-Cl
Na-Mg-Ca-Cl
830
730
Na-Ca-Cl-HCO3
2,200
Na-Mg-Cl
3,300
19,000
820
Quaternary
Limestone
Na-Cl
23,000
25,000
1,825
20,000
Tertiary
Sand
30,000
Na-Cl
Na-Ca-Cl-HCO3
530
Na-Cl
1570
Na-Ca-Cl
2435
Basement
0
200
400
600
800
1000
Distance (m)
Notes: Salinity in mg/L
Tertiary Clay

(a)

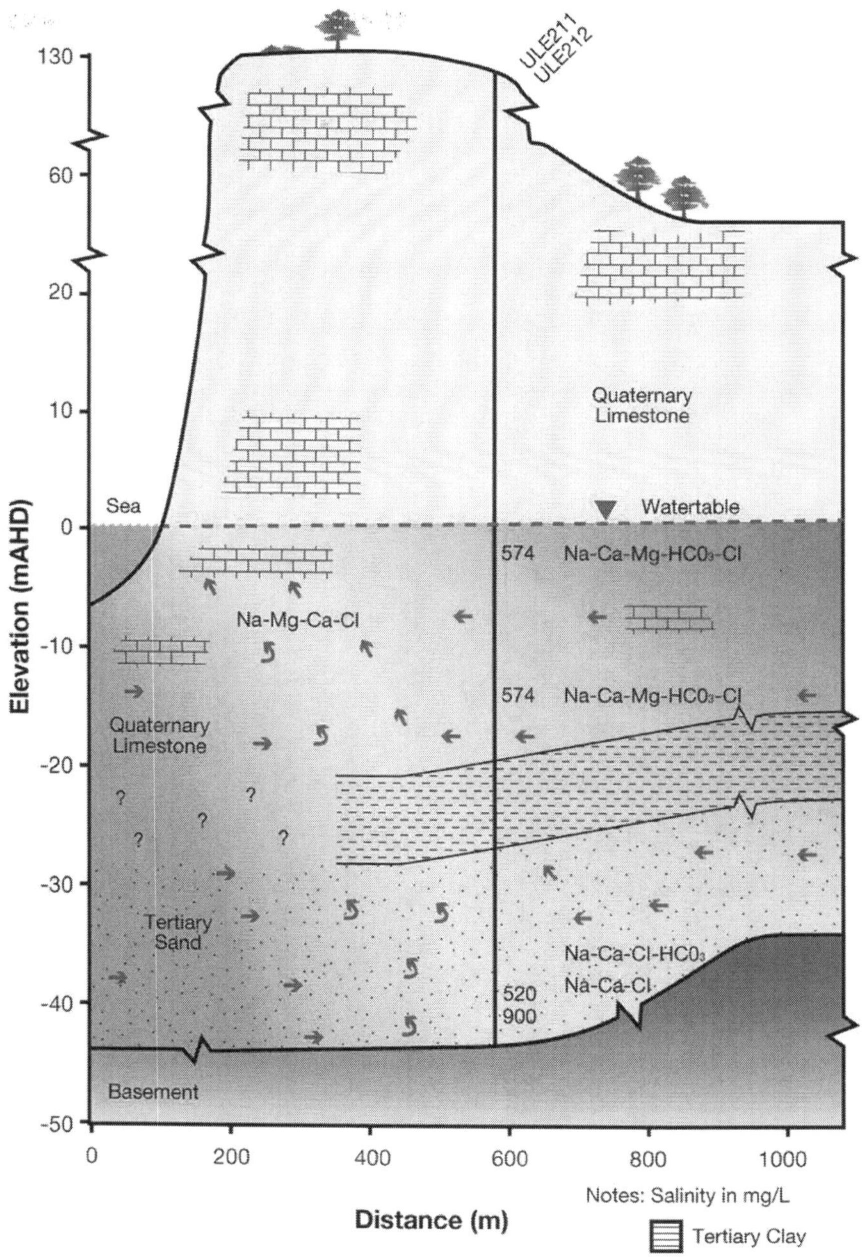

(b)

Figure 7. Uley South Saline Wedge. (a) Northern side; (b) Southern side.

their influence on groundwater chemistry. The geologic controls that led to these two distinct sources of different origin and age, were evidenced in the presence of two features, basement rock highs and absence of an aquitard.

Evaluation of hydrochemical data suggests that calcium carbonate dissolution, ion exchange processes and mixing with sea aerosols in the coastal zone are the main factors influencing groundwater chemistry. When the two groundwater sources with different ages and different chloride concentrations mix, results in unreliable recharge estimates are likely to result in using CMB and CFC methods. In addition, the subsurface basement rock high influences the currently observed seawater wedge. The toe of the saline wedge has been observed between 900 m and 500 m inland. Thus knowledge of geological control in the groundwater system, as well as the hydrogeochemical makeup of groundwater source(s), forms important parts of the characterisation of groundwater basin, ultimately contributing to effective management of the groundwater resources.

5. ACKNOWLEDGEMENTS

The editor and reviewers are thanked for evaluating this paper.

REFERENCES

1. L. Lakshmanan, R. Kanan and M. Senthi Kumar, "Major Ion Chemistry and Identification of Hydrogeochemical Processes of Groundwater in a Part of Kancheepuram District, Tamil Nadu, India," Environmental Geosciences, Vol. 10, No. 4, 2003, pp. 157-166.

2. M. Lorite-Herrera and R. Jiménez-Espinosa, "Impact of Agricultural Activity and Geologic Controls on Groundwater Quality of the Alluvial Aquifer of the Guadalquivir River (Province of Jaén, Spain): A Case Study," Environmental Geology, Vol. 54, No. 7, 2008, pp. 1391-1402.

3. R. M. Delinom, "Structural Geology Controls on Groundwater Flow: Lembang Fault Case Study, West Java, Indonesia," Hydrogeology Journal, Vol. 17, No. 4, 2009, pp 1011-1023.

4. G. Green, E. Watts, D. Alcoe, A. Costar, and L. Mortimer, "Groundwater Flow across Regional Scale Fault," Department for Water, South Australia, Technical Report DFW 2010/15, 2010.

5. S. T. Finch, A. Mccoy and E. Melis, "Geologic Controls on Groundwater Flow in the Mimbres Basin, South Western New Mexico," New Mexico Geological Society Guide Book, 59th Field Conference, 2008, pp. 189-198.

6. J. K. Mulwa, S. J. Garciri, J. O. Barongo, N. OpiyoAkech and G. K. Kianji, "Geological and Structural Control Influence on Groundwater Distribution and Flow in Ngong Area, Kenya," African Journal of Science and Technology, Science and Engineering Series, Vol. 6, No. 1, 2005, pp. 105-115.

7. T. T. Eaton, M. P. Anderson and K. R. Bradbury, "Fracture Control of Ground Water Flow and Water Chemistry in a Rock Aquitard," Groundwater, Vol. 45, No. 5, 2007, pp. 601-615.

8. T. E. S. Odeh, M. Schirmer and G. Strauch, "Structural Control of Groundwater Flow Regimes and Groundwater Chemistry along the Lower Reaches of the Zerka River, West Jordan, Using Remote Sensing, GIS and Field Methods," Vol. 58, No. 8, 2009, pp. 1707-1810.

9. V. F. Bensel, R. T. Van Balen and J. J. De Vries, "The Impact of Faults on the Hydrogeological Conditions in the Roer Valley Rift System: An overview," The Netherlands Journal of Geosciences/Geologie en Mijnbouw, Vol. 82, No. 1, 2003, pp. 41-54.

10. J. Meeder, "Karst Control of the Biscayne Aquifer Groundwater Flow Patterns. The Biscayne and Other Eogenetic Karst Aquifers: Characterization, Modeling, and Management," Geological Society of America, Annual Meeting and Exposition, Paper 77-6, 2012.

11. J. A. Cherry, B. L. Parker, K. R. Bradbury, T. T. Eaton, M. G. Gotkowitz, D. J. Hart and M. A. Borchardt, "Role of Aquitards in the Protection of Aquifers from Contamination: A State of the Science Report," AWWA Research Foundation, Denver, 2004.

12. R. W. Segnit, "Final Report on the Uley Wanilla Fresh Water Basin, Eyre Peninsula," South Australian Department of Mines, 1942.

13. R. K. Johns, "Geology and Mineral Resources of Southern Eyre Peninsula," Geological Survey of South Australia, Bulletin 37, 1961.

14. W. Morton and T. M. Steel, "Eyre Peninsula Groundwater Study-Uley South Basin, Progress Report 1, Southern Eyre Peninsula," Mineral Resources Reviews, South Australia, 128, 1970, pp. 33-48.

15. C. C. Wilson, "Geology of the Quaternary Bridgewater Formation of Southwest and Central South Australia," PhD Thesis, Flinders University of South Australia (unpublished), 1991.

16. R. G. Shepherd, "Report on Underground Water Investigations-Sleaford Bay-Coffin Bay Area," Department of Mines (Unpublished Report), 1963.

17. N. Harrington, D. Zulfic and D. Wohling, "Uley Basin Groundwater Modelling Project. Vol. 1: Project Overview and Conceptual Model Development," DWLBC Report 2006/01, Government of South Australia, 2006.

18. N. Somaratne, "Hydrogeology of the Uley South Basin," SA Water Hydrogeological Research, SA Water/2013-1, 2013.

19. S. L. Evans, "Estimating Long-Term Recharge to Thin, Unconfined Carbonate Aquifers Using Conventional and Environmental Isotopes Techniques: Eyre Peninsula, South Australia," Masters Thesis, Flinders University of South Australia (unpublished), 1997.

20. C. M. Ordens, A. D. Werner, V. E. A. Post, J. L. Hutson, C. T. Simmons and B. M. Irvine, "Groundwater Recharge to a

Sedimentary Aquifer in the Topographically Closed Uley South Basin, South Australia," Hydrogeology Journal, 2011.

21. G. S. Weissmann, Y. Zhang, E. M. LaBolle and G. E. Fogg, "Dispersion of Groundwater Age in an Alluvial Aquifer System," Water Resources Research, Vol. 38, No. 10, 2002, p. 1198.

22. L. N. Plummer, E. Busenberg and L. F. Han, "Chapter 5: CFCs in Binary Mixtures of Young and Old Groundwater. Use of Chloroflurocarbons in Hydrology: A Guide Book," International Atomic Energy Agency, 2006.

23. W. W. Wood and W. E. Sanford, "Chemical and Isotopic Methods for Quantifying Groundwater Recharge in a Regional Semi-Arid Environment," Ground Water, Vol. 33, No. 3, 1995, pp. 458-468.

24. W. W. Wood, "Use and Misuse of the Chloride Mass Balance Method in Estimating Groundwater Recharge," Groundwater, Vol. 37, No. 1, 1999.

25. E. Custodio, "Hydrogeochemistry and Traces," In: E. Custodio and G. A. Brugeman, Eds., Groundwater Problems in Coastal Areas, Studies and Reports in Hydrology, Vol. 45, 1987, pp. 213-269.

26. P. Heisig, "Groundwater Resources of the Clifton Park Area, Saratoga County, New York," US Geological Survey, Water Resources Investigation Report 01-4104, 2002.

27. W. R. I. HydroMetrics, "Seawater Intrusion Analysis Report, Seaside Basin, Monterey County, California," 2011.

CHAPTER 9

Multivariate Geostatistical Model for Groundwater Constituents in Texas

Faye Anderson

Texas A & M AgriLife Research Center at Beaumont, College Station, Texas, USA

ABSTRACT

Although many studies have explored the quality of Texas groundwater, very few have investigated the concurrent distributions of more than one pollutant, which provides insight on the temporal and spatial behavior of constituents within and between aquifers. The purpose of this research is to study the multivariate spatial patterns of seven health-related Texas groundwater constituents, which are calcium (Ca), chloride (Cl), nitrate (NO_3), sodium (Na), magnesium (Mg), sulfate (SO_4), and potassium (K). Data is extracted from Texas Water Development Board's database including nine years: 2000 through 2008. A multivariate geostatistical model was developed to examine the interactions between the constituents. The model had seven dependent varia- bles—one for each of the constituents, and five independent variables: altitude, latitude, longitude, major aquifer and water level. Exploratory analyses show that the data has no temporal patterns, but hold spatial patterns as well as intrinsic correlation. The intrinsic correlation allowed for the use of a Kronecker form for the covariance matrix. The model was validated with a split-sample. Estimates of iteratively re-weighted generalized least squares converged after four iterations. Matern covariance function estimates are zero nugget, practical range is 44 miles, 0.8340 variance and kappa was fixed at 2. To show that our assumptions are reasonable and the choice of the model is appropriate, we perform residual validation and universal kriging. Moreover, prediction maps for the seven constituents are estimated from new locations data. The results point to an alarmingly increasing levels of these constituents' concentrations, which calls for more intensive monitoring and groundwater management.

Keywords: Multivariate, Geostatistical, Groundwater, Constituents, Texas

1. INTRODUCTION

Groundwater is one of Texas natural resources that supplies the majority of the total water use in Texas [1] . Many local groundwater management districts are organized on county lines rather than on natural boundaries of the aquifers. Few counties take their water from more than one aquifer [2] . Groundwater quality covers physical, chemical, and biological aspects. Physical water qualities include temperature, turbidity, color, taste, and odor [3] . The increasing demand for high quality groundwater has driven many studies to investigate how the constituents' concentrations change over time and space, and what sources can be controlled in order to keep their levels within the acceptable range. Higher concentrations negatively affect the environment and public health. Examples of groundwater pollution sources include salt water intrusion, fertilizer leakage, natural erosions, and mine discharge.

This study focuses on the simultaneous spatial and temporal distributions of the seven most investigated groundwater constituents (five major and two minor constituents) over a nine-year period. A deeper understanding of the factors affecting constituents' levels may lead to a more successful and specialized programs designed at protecting groundwater from contamination. The results of this research project may be relevant to preventing and controlling groundwater contamination.

2. METHODS

2.1. Data

Data was obtained from the Texas Water Development Board Groundwater Database for all Texas wells from 2000 through 2008. The samples were checked for flagged values to ensure acceptable results in terms of reliable sampling, threshold conditions, or other criteria that can label the values as non-reliable. The wells are sampled periodically every four years (Figure 1). R version 2.9.0 was used. Descriptive statistics were performed using the build-in functions within the stats package. Variograms were plotted using geoR package.

2.2. Investigation of Temporal and Spatial Effects

To investigate temporal effects, Fisher's F-test was used to test the null hypothesis of non-changing variances between samples of the following years: 2000 versus 2004, 2001 versus 2005, 2002 versus 2006, 2003 versus 2007 and 2004 versus 2008. Fisher's F test p-values were all greater than the significance level of 0.05. Hence we cannot reject the null hypothesis of equal variances. Therefore, none of the constituents has shown any change in variance from the year 2000 to the year 2008. Repeated measures t-test was used to test the null hypothesis of non-changing means for the seven constituents (calcium, chloride, magnesium, potassium, nitrate, sulfate, sodium). The test was run to compare sample pairs of the following years: 2000 versus 2004, 2001 versus 2005, 2002 versus 2006, 2003 versus 2007 and 2004 versus 2008. It was found that none of the constituents has shown any change from the year 2000 to the year 2008. T-tests p-values were all greater than the significance level of 0.05. Hence the null hypothesis of equal means could not be rejected. Moreover, mapping of annual concentrations showed that for each of the seven constituents, the differences between two years levels was around zero. Based on the results for non-changing variances and means across the years, it was concluded that the data set does not contain temporal effects and all records between 2000 and 2008 were combined.

2.3. Exploratory Data Analyses

Exploratory analyses showed that Texas well depth means are similar all over the State, and that variances of constituents are lesser within an aquifer than between the aquifers. This means that the locality of a constituent has an effect on its level, which calls for a spatial model. Descriptive statistics showed high shifts of skewness and kurtosis from 0 and 3, respectively, which are the characteristic values of normal distribution. Therefore, the variables generally exhibit non symmetric distributions, with long tails and several outliers (Table 1). Because the data also had positive skewness as well as outliers, a log transformation on all the constituent data was performed. Furthermore, standard tests of univariate and multivariate normality (Shapiro-Wilk test and E-statistic (Energy)) [4] did not reject the null hypothesis of normality for the log-transformed data. Four records out of 3379 contained zero values. We neglected these records. Table 2 presents the pair-wise correlations between the transformed constituents across aquifers whereas Table 3 lists correlations within the Ogallala Aquifer, which are all non-negligible. Please note that the absence or drop of correlation between constituents does not imply the absence of spatial correlation.

A preliminary variogram analysis enables us to visualize the characterization of spatial correlation. At this point, we recognize that there may be mean effects that are unaccounted for. Nevertheless, these initial variogram estimates indicate consistent spatial behavior across aquifer. None of the constituents showed any significant anisotropy within an aquifer. Multivariate intrinsic correlation exists

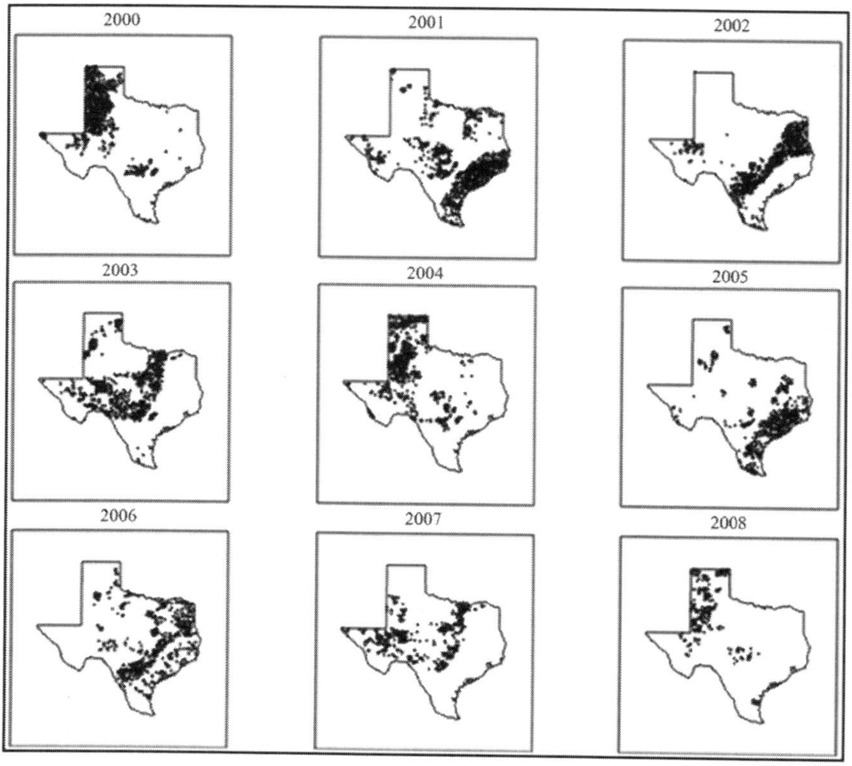

Figure 1. Groundwater four-year periodical sampling from 2000 to 2008.

Table 1. Across-aquifer descriptive statistics of the groundwater constituents.

Variable	Mean	Minimum	Maximum	Median	Std. Dev.	Skewness	Kurtosis
Calcium	93.72	0.35	1560.00	68.50	104.38	4.19	25.80
Chloride	159.32	1.85	17000.00	41.60	421.6111	18.41	635.51
Potassium	5.95	0.20	99.30	4.40	6.20	3.53	25.69
Magnesium	36.59	0.22	559.00	24.20	44.60	3.93	24.23
Sodium	122.51	2.39	13900.00	47.15	301.76	24.74	1052.38
Nitrate	17.33	0.00	425.88	8.01	31.68	5.03	35.69
Sulfate	192.40	0.00	5110.00	46.50	389.27	4.08	22.56

Table 2. Pair-wise inter-constituent correlations across aquifers.

Variables	ln(Ca)	ln(Mg)	ln(Na)	ln(K)	ln(SO_4)	ln(NO_3)	ln(Cl)
ln(Ca)	1	0.695	0.043	0.218	0.403	0.428	0.359
ln(Mg)	0.695	1	0.166	0.525	0.557	0.489	0.315
ln(Na)	0.043	0.166	1	0.628	0.696	0.026	0.832
ln(K)	0.218	0.525	0.628	1	0.629	0.206	0.541
ln(SO_4)	0.403	0.557	0.696	0.629	1	0.288	0.678
ln(NO_3)	0.428	0.489	0.026	0.206	0.288	1	0.157
ln(Cl)	0.359	0.315	0.832	0.541	0.678	0.157	1

Table 3. Pair-wise inter-constituent correlations within the Ogallala Aquifer.

	Ca	Mg	Na	K	SO_4	NO_3	Cl
Ca	1.00	0.67	0.64	0.44	0.76	0.52	0.81
Mg		1.00	0.72	0.77	0.87	0.46	0.82
Na			1.00	0.59	0.86	0.41	0.88
K				1.00	0.64	0.43	0.61
SO_4					1.00	0.45	0.81
NO_3						1.00	0.46
Cl							1.00

when the multivariate correlation of a multivariate data set is independent of the spatial correlation. Multivariate intrinsic correlation allows one to simplify data modeling. Under intrinsic correlation, the joint covariance matrix is given by the Kronecker product $\Sigma = V \otimes R$, where V is the standard variance-covariance matrix. We have $\Sigma^{-1} = V^{-1} \otimes R^{-1}$, which only requires the inversion of the 7×7 variance-covariance matrix V and the $n \times n$ spatial correlations matrix W [5] . One of the ways to test for intrinsic correlation is to calculate codispersion coefficients (i.e. the ratio of the cross versus the direct variograms). If the codispersion coefficients are constant, we can assume intrinsic correlation [5] [6] . The calculations of codispersion coefficients show that they are practically constant for all spatial scales. Their variances ranged from 0.01 to 0.04. This supports the intrinsic correlation hypothesis, i.e. a similar correlation matrix W holds at all spatial scales. In other words, when the covariance is structured according to the intrinsic correlation model, all direct and cross covariance functions are proportional to one basic covariance function. The model of intrinsic correlation is entirely specified by its spatial structure, and by the variance-cova- riance matrix [6] .

2.4. Principal Component Analysis

Principal components analysis (PCA) was conducted to study and visualize the correlations between the variables and hopefully be able to limit the number of variables to be measured afterwards, and to visualize observations in a 2- or 3-dimensional space in order to identify similarities and dissimilarities within observations. PCA was performed using the correlation matrix, which brings the measurements onto a common scale. In other words, a constituent which concentration varies between 0 and 1 will not weigh more in the projection than a constituent varying between 0 and 400. The Scree Plot of Figure 2 shows the cumulative variance explained by the seven principal components F1:F7. The principal components are sorted in decreasing order of variance, so the most important principal component is always listed first. Positions of the observations on a biplot are scores of the observations on the first two components. Distance between observations on this plane is "how close these observations are to each other" (Since we use two-dimensional plot we have ignored all remaining components). With respect to the first two principal components, nitrate is a bit isolated as compared to the other constituents. This suggests that nitrate levels might come from a different origin than the remaining constituents. The variation explained by each principal component does not exceed that of the first principal component which is 52%. The first two principal components explain 75% of the data variation. If further analysis needs to be performed on the 2-dimensional PCA scores instead of the seven-dimensional dataset, this leaves a significant amount of variation not accounted for. Therefore, principal components analysis results were not used to model the data.

3. RESULTS AND DISCUSSION

For our estimation we used the K-Bessel (Matern) model for the semivariogram model since its smoothness can be adjusted. Although the computation are cumbersome, the advantage of this model is that

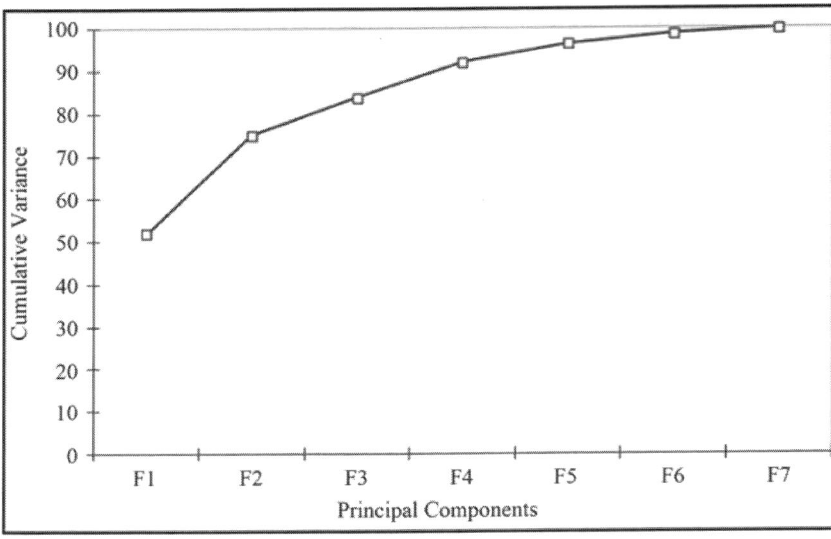

Figure 2. Scree plot.

the behavior of the semivariogram near the origin can be estimated from the data rather than assumed to be of a certain form. Also, changing the value of α, we can get other semivariogram models. For example, when $\alpha = 0.5$ we get the exponential covariance function [7] . We estimate β and $\Sigma(j)$ using iterative re-weighted generalized least squares IRWGLS [8] . For this model, the random functions are assumed to be second-order stationary. Isotropy of the data is also assumed. This means that the variogram is a function of the distance vector regardless of its direction. An alternative to IRWGLS would be maximum likelihood ML. We chose IRWGLS for edification purposes. The complexity of IRWGLS estimation lies in estimating the full covariance matrix of the data which is of a big size in our case. On the other hand, IRWGLS estimation is more efficient than weighted least squares and ordinary least squares. IRWGLS estimator of theta is efficient and has as little bias as possible [8] .

In order to determine the significant covariates to include in the model, we perform a model selection exercise. The research started with six covariates: altitude, latitude, longitude, major aquifer, water level and percentage of irrigated acres per county. The effect of percentage of irrigated acres per county is excluded from the analysis since it has shown no association with any of the constituents. Model selection was performed to decide on which of the remaining covariates to include in our model. The criteria for model selection is Akaike's information criterion AIC. One thousand records were selected based on simple random sampling without replacement. Many regressions were performed and their AIC values were compared. We started with no intercept. Then single predictor models, then two-predictor models and so on. The regression model with the least AIC value (3143.04) was found to be [Ca, Cl, NO_3, Na, Mg, SO_4, K] = longitude + altitude + water depth + aquifer effect (Table 4).

Figure 3 shows the empirical variogram of the OLS residuals. The seven variograms follow similar spatial behavior (range). Figure 4 shows the normalized residuals. The first seven standardized individual empirical variograms in Figure 4 look very similar to each other. The only difference between the variograms is the sill value. After running the IRWGLS algorithm for four iterations the results converge with tolerance less than 1e−05. Table 5 and Figure 5 show the Matern covariance function parameters after the fourth iteration. Practical range is about 44 miles, variance is 0.83, nugget is zero and smoothness parameter was fixed at 2. The range indicates that the maximal distance at which the constituents are spatially auto-correlated is 44 miles. Beyond 44 miles, the distance among wells does not affect the spatial structure of the data forming a sill of 0.83. At zero distance the variogram is zero. That is, there is no nugget effect. Comparing OLS residuals variance-covariance results showed that there is more correlation represented in the fourth iteration estimates, which is an indicator that IRWGLS is more efficient than OLS. To validate the model, one hundred locations were randomly selected from the original dataset. This dataset was not used in the estimation analysis. It was used to predict the constituent levels at these sites and study the behaviour of the residuals. Residuals were computed using the IRWGLS fourth iteration estimates. The scatter plots of the residuals do not follow a specific pattern. The residuals departure from the normality was negligible. Shapiro-Wilk test did not reject the null hypothesis of normality for the residuals. P-values were less than the significance level of 0.05. From the above discussion we conclude that the assumptions were reasonable and the choice of the model was appropriate.

Because the concentrations were log transformed prior to modeling (Table 6), the predicted concentrations on the prediction maps are contrasted to the log transformed values of EPA's maximum contaminant levels [9] (Figure 6). Prediction maps provide an easier tool for researchers and hydrologists to see which constituents' concentrations might exceed or go below the EPA tolerated levels. This also helps groundwater policy makers to act appropriately and in a timely manner in order to protect our natural resources. A grid of two

Table 4. Regression models AIC values.

Regression	AIC
Nitrate = 1	3643.62
Nitrate = Altitude	3530.97
Nitrate = Latitude	3573.52
Nitrate = Longitude	3507.15
...	...
Nitrate = Water Depth	3510.61
Nitrate = Longitude + Altitude	3509.11
Nitrate = Longitude + Latitude	3502.37
Nitrate = Longitude + Water Depth	3401.39
...	...
Nitrate = Altitude + Latitude + Longitude + Water Depth	3371.91
...	...

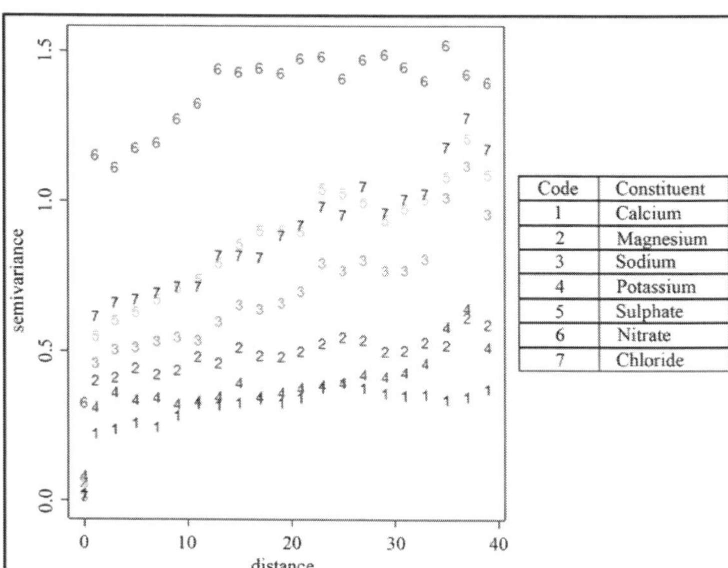

Figure 3. OLS residuals empirical variogram.

hundred and sixteen randomly chosen locations was formed. Elevations of points were estimated using global positioning system website GPSies. Table 7 presents the constituents with predicted concentrations that exceed the EPA's maximum concentration levels.

Table 5. Matern covariance function parameters after fourth iteration, Kappa = 2.

Nugget (tausq) = 0
Range (miles) = 8.22
Sill (sigmasq) = 0.83
Kappa (smoothness parameter) = 2
Practical Range = 44.12 (miles)

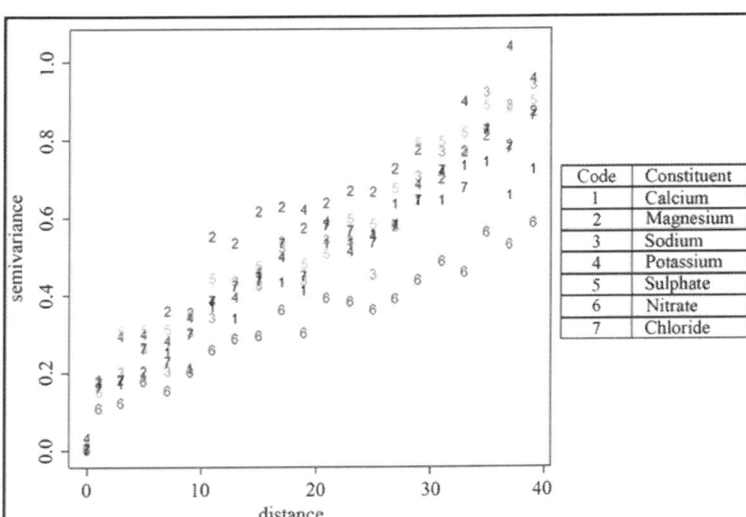

Figure 4. OLS standardized residuals empirical variogram.

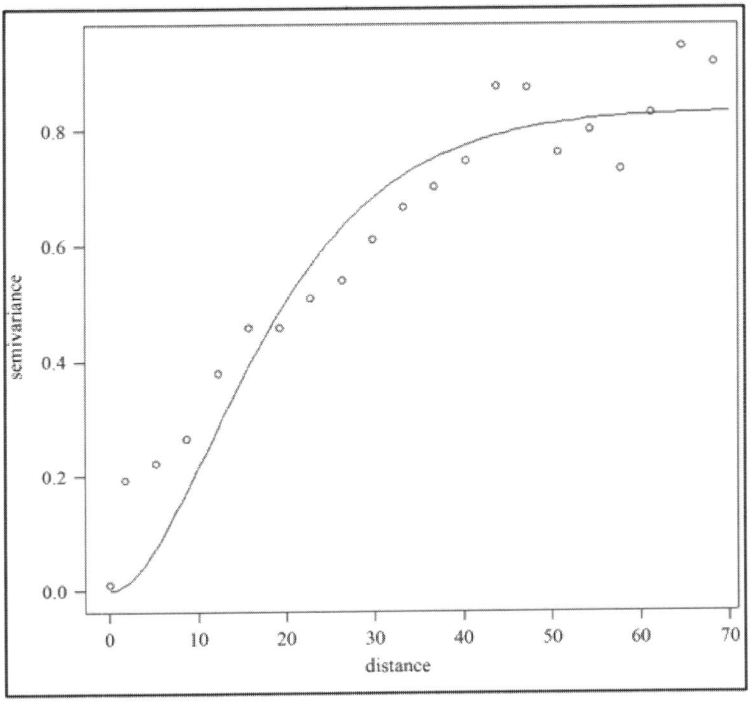

Figure 5. Fourth IRWGLS iteration matern function estimate variogram.

Table 6. EPA tolerated concentrations.

Constituent	Mx: The maximum level allowed in public water supplies by the EPA	Log(Mx)
Calcium	And magnesium 200 mg/L	5.298317
Chloride	250 mg/L	5.521461
Magnesium	And calcium 200 mg/L	5.298317
Nitrate	10 mg/L	2.302585
Potassium	10 mg/L	2.302585
Sodium	20 mg/L	2.995732
Sulfate	250 mg/L	5.521461

Table 7. Constituents with predicted concentrations higher than MCLs.

Constituent	Affected aquifers
Calcium	Gulf Coast, Ogallala, Edwards-Trinity Plateau and Seymour
Nitrate	Ogallala, Seymour and Edwards-Trinity Plateau
Chloride	Ogallala Edwards-Trinity Plateau and Hueco-Mesilla Bolson
Potassium	Ogallala
Sodium	All
Sulfate	Ogallala and Edwards-Trinity Plateau

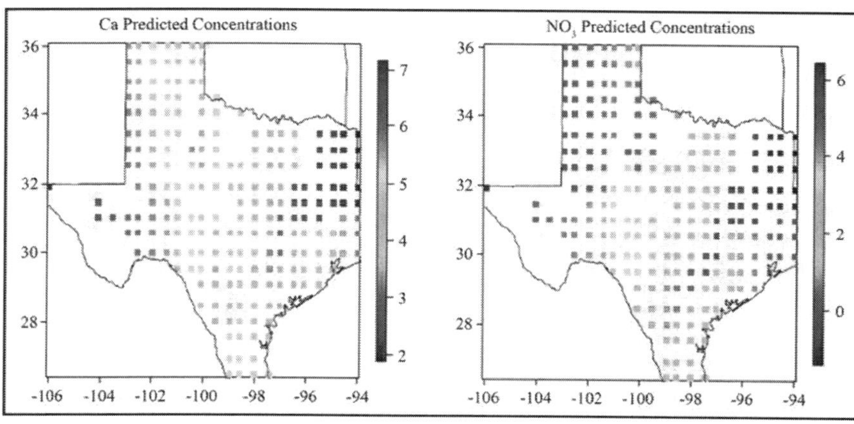

Figure 6. Predicted concentrations for five constituents in lnMg.

4. CONCLUSION

This study focused on seven of the most researched groundwater constitutents in Texas. Namely, calcium (Ca), Faye Anderson (Cl), nitrate (NO_3), sodium (Na), magnesium (Mg), sulfate (SO_4), and potassium (K). Data were extracted from TWDB database for the nine years from 2000 to 2008. No temporal effects were found in the concentrations but data was found to be intrinsically correlated. This allowed to spatially model the between-con- stituent effects individually using IRWGLS. The multivariate geostatistical model used latitude, longitude, elevation, water depth and aquifer effect as the mean covariates. Also, a covariance structure was estimated to link the constituents' concentrations with the spatial structure. The non-temporally different concentrations of the seven constituents points to the value of geostatistical modeling techniques as a valuable estimation method. This study provides an example where simultaneous modeling of pollutants can be conducted based on locality, which closes a big gap in environmental research where interaction between variables is significant. This study also shows that calcium, chloride, nitrate, sodium, magnesium, sulfate, and potassium follow similar increasing trends. The high concentrations that exceed EPA tolerated levels presented in the prediction maps (Figure 6) agree with recent research observations [10] , which further validates the model as a tool for hydrologists and policy makers to accurately estimate the seven constituents simultaneously based on location.

REFERENCES

1. Texas Water Development Board (TWDB) (2007) Water for Texas 2007. Volume II, p. 161.

2. Sansom, A., Armitano, E.R. and Wassenich, T. (2008) Water in Texas: An Introduction. UT Press.

3. Westlake, K. (1995) Landfill Waste Pollution and Control. Horwood Publishing.

4. Szekely, G.J. and Rizzo, M.L. (2013) Energy Statistics: A Class of Statistics Based on Distances. Journal of Statistical Planning and Inference, 143, 1249-1272.

5. Piegorsch, W.W. and El-Sharaawi, A.H. (2002) Encyclopaedia of Environmentrics. John Wiley and Sons, Hoboken, 1368.

6. Wackernagel, H. (2003) Multivariate Geostatistics: An Introduction with Applications. 3rd Edition, Springer.

7. Waller, L.A. and Gotway, C.A. (2004) Applied Spatial Statistics for Public Health Data. Wiley-IEEE. [Paper reference:1]

8. Schabenberger, O. and Gotway, C.A. (2005) Statistical Methods for Spatial Data Analysis. Chapman & Hall/CRC Press.

9. United States Environmental Protection Agency (EPA) (2014) Drinking Water Contaminants.

10. Chaudhuri, S. and Ale, S. (2014) Long Term (1960-2010) Trends in Groundwater Contamination and Salinization in the Ogallala Aquifer in Texas. Journal of Hydrology, 513, 376-390.

10

CHAPTER

Field Measurement of Hydraulic Conductivity of Rocks

Maria Clementina Caputo and Lorenzo De Carlo

Water Research Institute (IRSA), National Research Council (CNR Via F. De Blasio, Bari, Italy

INTRODUCTION

Groundwater often represents the main and most precious source of drinking water supply for the population. In recent decades, overexploitation, uncontrolled anthropogenic actions and continuous reduction of rainfall, due to climate change, led to a depletion of the water resource by affecting both its quantity and quality. The scientific community pays attention to this particular environmental issue in order to implement effective strategies for the safeguarding, protection, and remediation of the aquifers. Over the last decade, considerable efforts have been made to develop methodologies and techniques aimed at improving knowledge of the processes that are based in the unsaturated zone or vadose zone—the portion of the soil above the groundwater level—within which flow and transport processes occur. In this portion of subsoil, in fact, important physicochemical phenomena take place, which regulate the environmental balance of the hydrogeological system, such as the ability to store water and transport it into the ground below. It also has a natural protective function as a filter for any potential pollutants carried by fluids circulating in the solid matrix before reaching the groundwater. The knowledge and the understanding of the processes that take place in the unsaturated zone are, therefore, essential for groundwater management and protection, to evaluate the recharge rate and assess groundwater vulnerability. In particular, the measurement and monitoring of the unsaturated hydraulic properties are very important, even though it is difficult and expensive (Castiglione et al., 2005). The methods and techniques developed are designed to investigate the unsaturated flow process in the soils. When the vadose zone consists of rock, rather than soil, technical aspects increase the difficulties in several ways (Bogena et al., 2007; Kizito et al., 2008). Usually, different kinds of probes are used to monitor the water infiltration in soils: the Time Domain Reflectometry (TDR) (Jones et al., 2004; Robinson et al., 2003), the Frequency Domain Reflectometry (FDR) and multi-sensor capacitance probes (Baumhardt et al., 2000; Seyfried et al., 2004) are used to measure the water content in the subsurface, while tensiometers measure water pressure (Masbrucha & Ferré, 2003). These devices are hard to utilize in the rocks, mainly because the probes are brittle. Therefore it is difficult to install them, as there needs to be good contact between the rock and the sensor to reduce the uncertainty of the measurements due to the gap effects. Field studies, set up to measure hydraulic conductivity, have employed infiltrometer tests under different conditions, but they have rarely been performed directly on the outcropped rock, owing to the difficulty of installation. Reynolds et al. (2002) conducted infiltrometer tests under different conditions, Ledds-Harrison & Youngs (1994) used very small diameter rings (from 1.45 mm to 2.5 mm) for field measurements on individual soil aggregates, Youngs et al. (1996) used a 20 m diameter infiltrometer cylinder to measure highly structured and variable materials that could not be sampled adequately by a smaller cylinder. Castiglione et al. (2005) developed in the laboratory a tension infiltrometer ring, 4 cm in height and 27.5 cm in diameter, suitable for accurate measurements of infiltration into a big sample of fractured volcanic tuff, at very low flow rates over long equilibration times. Most field studies employ cylinder infiltrometers with diameters ranging commonly from 1 to 50 cm, which are poorly representative of the heterogeneous media, such as fractured rocks, in which hydraulically important fractures may, typically, be spaced further apart than the cylinder's diameter. Indeed, up to now infiltrometer tests have rarely been performed directly on-site on rock outcrops. This chapter describes a methodology to obtain the field-saturated hydraulic conductivity, Kfs, by using a ring infiltrometer, with a large (~2 m) adjustable diameter, developed for measuring quasi-steady infiltration rates on outcropped rock. Kfs is the hydraulic conductivity of the medium (soil or rock) when it has been brought to a near-saturated state by water applied abundantly at the land surface, typically by processes such as ponded infiltration, copious rainfall or irrigation. The proposed device is inexpensive and simple to implement, as well as very versatile, owing to its large adjustable diameter that can be fixed on-site. Moreover, certain practical problems, related to the installation of the cylindrical ring on the rock surface, were solved in order to achieve a continuous and impermeable joint surface between the rock and the ring wall. An issue of major concern is linked to the edge effects, related to the radial spreading of the infiltrating water; obviously smaller rings are more influenced by these effects. Swartzendruber & Olson (1961) and Lai

& Ren (2007) found that the ring infiltrometer needs a diameter greater than 1.2 m and 0.8 m, respectively, to avoid the edge effects. For this reason, the proposed large ring infiltrometer is made of a strip of flexible material with which build the cylinder on-site, with a suitable diameter in relation to the lithological and topographical features of the field. The flexible material, such as plastic or glass resin, allows the minimization of the size of the ring and, therefore, its movement easily, in order to acquire a large number of independent Kfs measurements over a given area. In fact, because of the extreme spatial variability of Kfs, its value finds statistical consistency in multiple tests. Geophysical techniques were coupled with the infiltrometer tests in order to monitor, qualitatively, the water infiltration depth, to allow a rapid visualization of the change in water content in subsurface and to ensure that the decrease in the water level in the ring was caused mainly by vertical water infiltration, and not by the lateral diversion of water flow. Since the late 1980s, many geophysical applications have been aimed at hydrogeological studies. White (1988) conducted electrical prospecting to determine the direction and the flow rate of saline aquifers using a tracer. Daily et al. (1992) used borehole electrical resistivity surveys to obtain the distribution of electrical resistivity in the subsurface, and compare the results with infiltrometer tests. Recently, electrical resistivity techniques have been used to monitor hydrogeological processes. Cassiani et al. (2006) conducted a monitoring test of a salt tracer by means of the application of electrical resistivity tomography using a time-lapse technique. The movement of the tracer was monitored with geophysical images. The methodology described in this chapter was carried out on different lithotypes in order to verify the applicability of the experimental apparatus in very different geological conditions. In particular two cases are described: the Altamura test site that represents a case of hard sedimentary rock consisting of limestone, and the San Pancrazio test site as an example of a soft porous rock, specifically calcarenite. Both sites, located in the Puglia region of southern Italy, have been studied because they are both contaminated areas, for which the understanding of the flow rate in the subsurface is very important in order to design a remediation strategy. The experimental approach gave good results in both situations that mean it could be used successfully on each different lithology of outcropped rock.

METHOD

The objective of this chapter is to describe a methodology to obtain the field-saturated hydraulic conductivity, using a large ring infiltrometer installed directly on the outcropped rock. The method considers an integrated approach that combines the infiltrometer test with geophysical techniques, in order to visualize the change in water content in the subsurface during the experiment. Among geophysical techniques, we have used electrical resistivity images (ERI) because the electrical resistivity is extremely sensitive to subsurface water content. Moreover, electrical time-lapse resistivity measurements (Binley et al., 2002; Deiana et al., 2007), carried out simultaneously with the infiltrometer test, allow us to check the performance of the experimental apparatus and, particularly, that the decrease of water level in the ring is due to the infiltration and not to the losses along the edge of the ring.

Large ring infiltrometer

A ring infiltrometer for studying water flow in rock formations needs to be made from a tough material, suitable for installation on the rock surface. At the same time, the material must be both flexible, to allow a cylinder to be built that is adjustable to the field conditions, and light, to facilitate transport and setup. Where the ground surface is very irregular, small adjustments to the ring diameter ensure efficient sealing, thus facilitating the setting-up of the device. If electrical methods are to be used for the detection of water content or salinity, the ring must also be made of a non-metallic material. The large ring infiltrometer was, therefore, designed in a light and flexible material, such as plastic or glass resin, to allow easy installation on the rock surface, and the adaptation of its size to the field conditions. Strips of the flexible material 30 cm wide and 0.2 cm thick were used to build in-situ rings of internal diameter of about 2 m, by sealing the two ends of the strip with impermeable adhesive tape to form the cylinder. The large diameter provides infiltration data that are representative both of the anisotropy and the heterogeneity of the rock, characteristics which cannot be sampled adequately by using smaller rings. It thus allows an improved representation of the natural system's heterogeneity, while also taking into consideration irregularities in the soil/rock subsurface. In addition, the large diameter allows the avoidance of the edge effects that usually occur during infiltration, such as the radial spreading of the infiltration water. Among a variety of materials tested to seal the ring to the ground, both in the laboratory and in the field, gypsum was shown to be the most suitable because it is cheap, easy to prepare in situ and to apply on the ground in a previously hollowed furrow, ensuring good sealing. Clay, instead, needed to be worked for a long time in order to obtain a consistency that would allow the ring to be sealed to the ground. It nevertheless gave poor results, because the clay did not grip the ring wall firmly on the ground. Moreover, tests carried out with clay at the laboratory showed water losses from cracks in the clay due to its characteristics of swelling and shrinking. Silicon was also tested, although using it for sealing a large-diameter infiltrometer would be expensive. However, it was also discarded, as it did not seal well due to the dissimilarity of the soil, rock, and plastic materials being sealed. Polyurethane foam was easy to apply and proved impermeable, but was also discarded because it was expensive and did not grip well on the rock and/or soil.

Geophysical techniques

Geophysical techniques are used routinely for subsurface investigation, as they provide information about the physical properties related to geological and hydrogeological conditions. There are two main groups of geophysical methods: the active methods, which measure the subsurface response to electromagnetic, electrical, and seismic artificially generated signals; and the passive methods, which measure the earth's natural fields, such as the magnetic, electrical, and gravitational ones. Among these, the electrical resistivity method is a relatively new imaging tool in geophysics; it is an active method that determines the subsurface distribution of electrical resistivity from a large number of resistance measurements. The electrical resistivity, ρ, is an intrinsic property of rock and soil, and is a measure of how strongly a material opposes the flow of electric current. The direct current resistivity tomography method considers spatial variations in electrical resistivity among geologic materials for mapping the subsurface structures. It is used extensively in the search for suitable groundwater sources, and also to monitor types of groundwater pollution. In engineering it is used to locate subsurface cavities, faults and fissures, permafrost, mineshafts etc., and in archaeology, for mapping out the areal extent of the remnants of buried foundations of ancient buildings, and many other applications (Reynolds, 1998). The electrical resistivity method is based on the Ohm's laws; the first of which states that when the electrical current, I (ampere, symbol A), passes through an electrically uniform medium of side length, L (m), the material has a resistance, R (ohm, symbol ⍰⍰, resulting in a potential drop, V (volt, symbol V), between opposite faces of the medium. The first Ohm's law correlates the three physical parameters into the formula:

$$V = RI \tag{1}$$

The resistance, R, is proportional to the length, L, of the resistive material and inversely proportional to the cross-sectional area, S (m^2) (second Ohm's law)

$$\rho = \frac{RS}{L} \tag{2}$$

where ρ is the electrical resistivity (ohm x m, symbol ⍰m); it is the inverse of the conductivity (siemens/m, symbol S/m).

Electrical current can flow in rocks and soils by three main mechanisms: 1) electrolytic conduction that occurs by means slow migration of ions in a fluid electrolyte, controlled by ions type, ion concentration, and ionic mobility; 2) electronic conduction that occurs in metals through rapid movement of electrons; 3) dielectric conduction that occurs in conducting materials, or insulators, in the presence of an external alternating current when atomic electrons are shifted slightly relative to the nucleus. In most rocks, electrical current flows by electrolytic conduction (Telford et al., 1990). The electrical resistivity of the soils and rocks correlates with other soil/rock properties which are of interest to the geologist, hydrogeologist, and geotechnical engineer. Many factors affect electrical resistivity, such as texture, porosity, water saturation, clay content, permeability and temperature. Thus, the resistivity measurements cannot be related directly to the type of soil or rock in the subsurface without direct sampling, or some other geophysical or geotechnical information. Porosity is the major controlling factor of the changing resistivity of the rock, because the electricity flows in the near surface by means of the passage of ions through the pore space of the subsurface materials. The porosity (amount of pore space), the permeability (connectivity of pores), the water (or other fluid) content in the pores, and the presence of salts all are contributing factors to changing resistivity. Archie (1942) developed the empirical formula of the effective resistivity of the rock, ρe:

$$\rho_e = \rho_w a \varphi^{-m} S^{-k} \tag{3}$$

where φ (%) is the porosity, S (adimensional) is the volume fraction of pores with water, ρw is resistivity of pore fluid, a, m and k are physical parameters. ρw is influenced by dissolved

salts and can vary between 0.05 ?m for saline groundwater, up to 1,000 melt water. Archie's law ignores the effect of pore geometry, but it is a reasonable approximation for many sedimentary rocks.

2.2.1 Electrical resistivity imaging

The technique of electrical resistivity imaging utilizes measurements of electrical potential associated with the subsurface electrical current flow, generated by a direct current. In the electrical resistivity method, the spatial variation of resistivity in the field is determined using four electrode measurements, and is based on measuring the potential difference between one electrode pair while another electrode pair, used as the current source, transmits the current (Dahlin, 2001). Measurements of the potential difference between the two electrodes allow the determination of the apparent resistivity, �a:

$$\rho_a = \frac{K\Delta V}{I} \quad (4)$$

where K (m) is a geometric factor depending on the used array. The measured quantity is called apparent resistivity because the resistivity values measured are averages over the total current path length, but are plotted at one point for each potential electrode pair. The data can be arranged in a 2-D plot, called pseudosection, which displays both horizontal and vertical variations in apparent resistivity. The conventional presentation places each measured value at the intersection of two 45-degree lines through the centres of the dipoles (Edwards, 1977). The interpretation of the resistivity data consists of two steps: a physical interpretation of the measured data, by means of an inversion process that results in a physical model; and a geological interpretation of the resulting physical parameters (Dahlin, 1996). Many configurations of array can be used for measuring the distribution of electrical resistivity of the subsurface. The arrays most commonly used for 2-D imaging surveys are: Wenner, dipole-dipole, Wenner-Schlumberger. The choice of the best array considered for the surveys, depends on the depth of investigation, the sensitivity of the array to vertical and horizontal changes in the subsurface resistivity, the horizontal data coverage, and the signal strength (Loke, 2001). In general, Wenner array is relatively sensitive to vertical changes in the subsurface resistivity, but it is less sensitive to its horizontal changes. It is useful for recognized horizontal structures (vertical resistivity changes), but relatively poor in detecting narrow vertical structures (horizontal resistivity changes). Dipole-dipole is good for mapping vertical structures, such as dykes and cavities, but relatively poor for mapping the horizontal structures such as sedimentary layers or water table. Wenner-Schlumberger is moderately sensitive to both horizontal and vertical structures. In areas where both types of geological structures are expected, this array might be a good compromise between the Wenner and the dipole- dipole array.

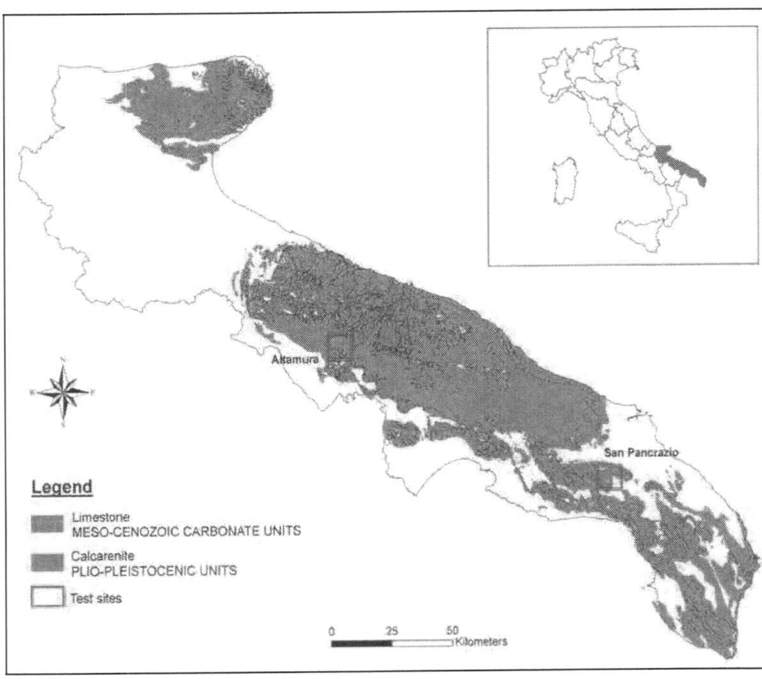

Fig. 1. Schematic geological map of the Puglia region.

FIELD TESTS

The infiltration experiments described in the chapter refer to two sites that differ in their geological and hydrogeological conditions, in order to show how the methodology can be applied widely. The sites consist of outcrop of hard sedimentary rock, fractured limestone, in the case of Altamura, and soft sedimentary rock, in the case of San Pancrazio (Fig. 1). The different lithology implies different infiltration process and, consequently, different experimental approaches and interpretations of the experimental data. All these aspects are described in detail for each test site.

Test on limestone

The infiltration experiments were performed in the study area belonging to Altamura territory, located in southern Italy (Fig. 1). The site is part of the Murge plateau, formed from Cretaceous limestone during the Paleogene and Neogene periods. The limestone is part of a sedimentary sequence of a carbonatic platform some thousands (up to 3,000) of meters thick. The main rock units consist of calcilutites, which in the study area constitute a unit known as *"Calcare di Altamura"* and *"Calcare di Bari"* (Ciaranfi et al. 1988). During a large part of the Neogene, the area was uplifted and affected by tectonic stress and karst morphogenesis. The experimental tests were conducted on the top of a karstic fractured limestone formation constituting the deep Murge aquifer, which has a water table ranging in depth from 400 to 500 m below the ground surface. The fractured limestone studied is characterized by low effective matrix porosity with respect to fracture porosity. Borgia et al. (2002) measured the effective matrix porosity of the limestone, at field scale, and found values up to 0.1%, together with a permeability lower than 24 mD (millidarcy), and a hydraulic conductivity of 2.1×10^{-2} m d^{-1} (for water at 20°C), on average. These values indicate a conductivity 1,000 times lower than that of the fracture network (8–77 m d^{-1}) of the same limestone formation (Masciopinto, 2005), which is characterized by an effective fracture porosity at field scale of up to 0.4% (Borgia et al., 2002). In light of these results, it is clear that the water flow rate component in the rock matrix can be considered negligible as compared to the water flow rate component in the network of fractures. This site was chosen for testing the newly developed large ring infiltrometers because it was affected by untreated sludge waste deposits that have caused soil and rock contamination by heavy metals (chromium, lead, zinc, nickel and cadmium), as well as hydrocarbons. Several soil chemical analyses carried out, showed chromium and nickel contamination, with concentrations higher than both Italian and European legal limits. The infiltrometer tests were carried out at two locations, 300 m apart, to evaluate the quasi-steady vertical flow rates in outcrops of limestone that showed different fracture frequencies and sizes on visual inspection. Specifically, the site of test #1 showed prominent visible fractures of about 5 cm spacing and about 1 mm aperture, whereas the site of test #2 showed no visible fractures (Fig. 2). A 2 m diameter ring, coupled with an external ring, was used in test #1. In this case, the infiltration area consisted of a central part with numerous visible fractures (Fig. 2a), while the rest was partially covered by soil, 5–10 cm thick, on average. For this reason, at this site, a second ring with a larger diameter (2.2 m) was positioned (Fig. 2b), in order to improve the hydraulic packing during the water infiltration test. Continuity between the ring infiltrometer wall and the rock-soil surface was obtained by filling the space between the external and internal rings with gypsum, up to a height of about 2 cm, to create a seal.

As a result, the second ring improved the gypsum sealing of the first ring to the ground, overcoming the challenge presented by the presence of different media (rock and soil) along the edge of the ring. In the second test (#2), located in an area with an outcrop of limestone characterized by no visible fractures, a 1.8 m diameter ring was inserted into a thin furrow 2 cm deep, previously hollowed into the rock (Fig. 3), which was then also sealed with gypsum. In both tests, the large diameter ring allowed the obtaining of data more representative of the anisotropy and heterogeneity of rocks than those obtained with a small ring. The difference in size of the rings used in the two tests was due simply to the local field conditions; in fact, the device is designed to allow the diameter to be adjusted easily when the irregularity of the surface at the test site so requires. In both cases the infiltration tests have been supported by ERI, carried out simultaneously by means "time-lapse" technique, in order to monitor the infiltration and redistribution of water in the subsurface. The water used for the infiltration test had moderate salinity (electrical conductivity of 2.39 mS/cm) to enhance the subsurface electrical resistivity measurements.

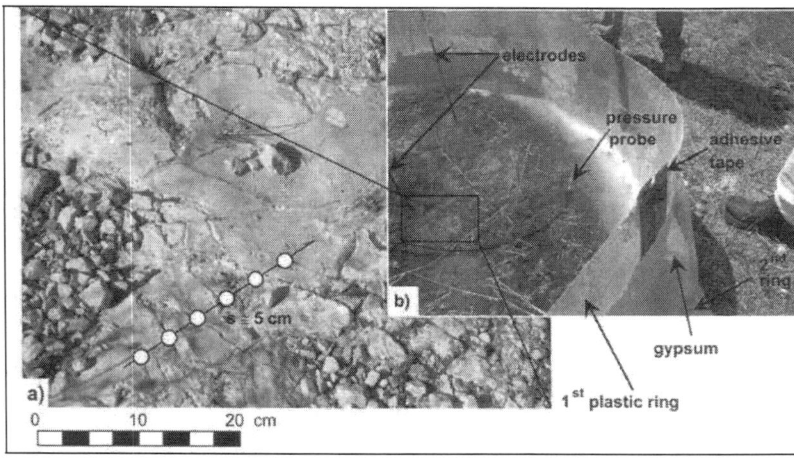

Fig. 2. Test # 1 at the Altamura site: a) fractured limestone with visible fractures of about 5 cm of spacing (s); b) plastic ring infiltrometer resting directly on the ground and sealed with gypsum by using a second ring.

In test #1, the resistivity data were collected using a 7.5 m long straight-line array of 16 steel electrodes with 0.5 m spacing, oriented in the parallel direction to the visible fractures. In test #2, two similar linear arrays were set up, perpendicular to each other, with the same length and electrode spacing, crossing at the center of the ring infiltrometer (Fig. 4.). The Wenner array was chosen over other array types because it provides a good signal-to-noise-ratio. Additionally, it is also highly sensitive to vertical changes in the subsurface resistivity. This makes the Wenner array a useful tool in studying the movement of the wetting front in time. The investigation depth is about 1.5 m, easily achievable with the chosen electrode configuration.

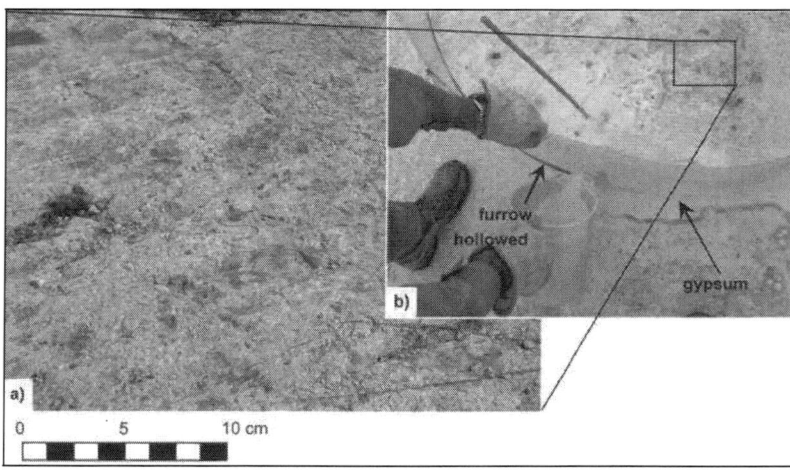

Fig. 3. Test # 2 at the Altamura site: a) limestone without visible fractures; b) plastic ring infiltrometer installed into a thin (2 cm) furrow hollowed in the limestone and sealed with gypsum.

Both tests used the IRIS–SYSCAL Pro Switch 48 instrument to acquire electrical resistivity measurements. To obtain 2-D resistivity models, the field data were inverted using Res2Dinv software (Griffiths & Barker, 1993; Loke & Barker, 1996). In the processing of resistivity data, an inversion routine based on the smoothness-constrained least-squares method was implemented (deGroot-Hedlin & Constable, 1990). The 2-D model used by the inversion program divides the subsurface into a number of rectangular blocks, whose arrangement is linked to the distribution of points in pseudosections. The distribution and size of the blocks is generated automatically by the program, using the distribution of the data points as a rough guide. The depth of the bottom row of blocks is set to be approximately equal to the equivalent depth of investigation (Edwards, 1977) of the data points. The optimization method basically tries to reduce the difference between the calculated and measured apparent resistivity values, by adjusting the resistivity of the model blocks. A forward modeling subroutine is used to calculate the apparent resistivity values, and a non-linear least-squares optimization technique is used for the inversion routine (Loke & Barker, 1996). A measure of this difference is given by the root-mean squared (RMS) error. However, the model with the lowest possible RMS error can sometimes show large and unrealistic variations in the model resistivity values, and might not always be the "best" model from a geological perspective. The water in the ring reached a maximum level of 0.13 m above the ground surface in both tests. During approximately 2 hours of the falling-head infiltrometer tests, the specific lateral water leakage fluxes were about 8% (0.1 m d^{-1}) and 3% (0.007 m d^{-1}) of the infiltrate water fluxes in the rings, for the first and second tests, respectively. The specific (i.e., per unit of ring area) leakage rates, due

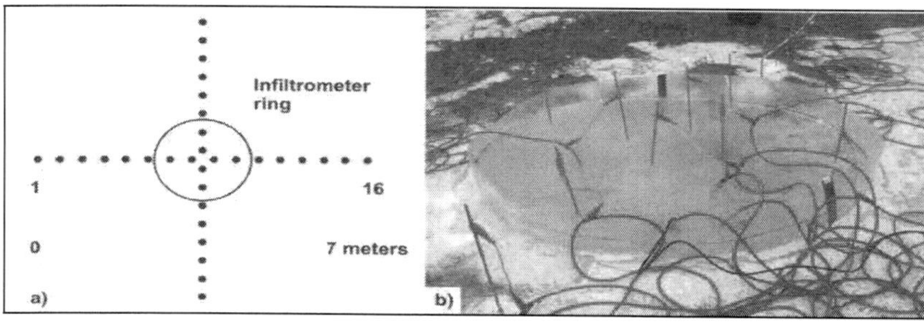

Fig. 4. Test # 2 at the Altamura site: a) scheme of electrode array; b) detail of field installation of electrodes.

not to the water flux from the edge of the ring (i.e., losses) but to the amount of infiltrated water which flowed laterally within the subsurface, was estimated by dividing the total water volume of the lateral diverted water flow by the duration of the infiltration (2 and 3.2 hours for tests #1 and #2, respectively). This estimation was performed on the basis of the areas of the wetted surfaces (1.6 and 1.5 m2 for tests 1# and 2#, respectively), which were observed around the rings at the end of each test, and multiplying them by the average thickness of the wetted layers (5 and 3 cm, respectively), and by their porosities (0.29 and 0.03) (Borgia et al. 2002). The thickness of the wetted layers was derived from measurements on core samples drilled at the test sites. During the experiments, water levels in the ring infiltrometer were monitored using a submersible pressure probe (PTX DRUCK

LTD).

Test on calcarenite

The study area is an abandoned calcarenite open quarry, located near San Pancrazio city (Fig. 1). The quarry was used for about ten years, from 1980 to 1990, for the disposal of waste from a pharmaceutical company that produced antibiotics (in particular erythromycin), using fermentative processes and subsequent chemical transformations causing subsurface contamination. For this reason, field infiltrometer tests, coupled with electrical resistivity measurements, were carried out to evaluate the flow rate of potential contaminants in the porous aquifer. The geology of the study area consists of Cretaceous bedrock formed from dolomitic limestone and limestone, overlaid unconformably by Plio- Pleistocenic calcarenite. The oldest formations contain a deep and confined aquifer characterized by a potentiometric surface, ranging from 70 m to 80 m, in depth below ground surface. Electrical resistivity measurements have detected a shallow aquifer with a water table about 25 m below ground surface, and about 12 m below the bottom of the quarry in the Plio-Pleistocenic calcarenite. A large infiltrometer ring was installed in the field directly on the rock surface with an adaptable size to fit the condition in the field. A strip of 30 cm high flexible plastic material was used to build an in-situ infiltrometer ring, of about 2 m in diameter, sealing the two edges with impermeable tape. The ring was installed into a 5 cm deep thin furrow hollowed in the calcarenite at the bottom of the quarry, and sealed with gypsum (Fig. 5).

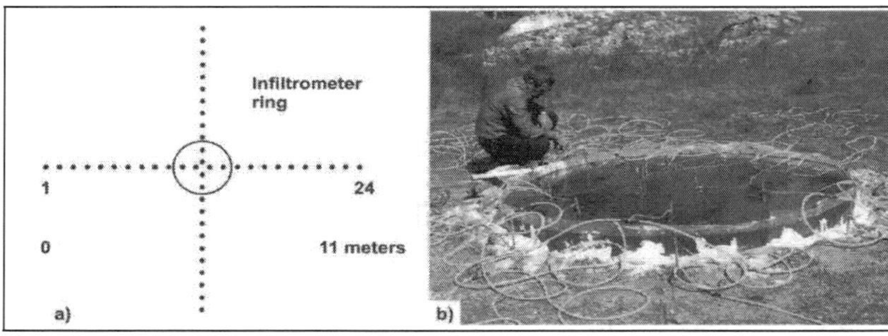

Fig. 5. Test at the San Pancrazio site: a) scheme of electrode array; b) detail of field installation of electrodes.

The infiltration tests were performed with constant and falling head conditions. About 0.27 m^3 of water was poured in the ring until the water level reached 8 cm from the bottom, and then another 0.2 m^3 to maintain that constant level in the ring for the duration of the test, which took 3 hours. No more water was added for the falling-head test because it started from the constant water level considered for the first test (under constant head condition). The second test took 2.5 hours. In order to study the infiltration/redistribution of water in the subsurface, electrical resistivity tomography was carried out simultaneously with the infiltrometer tests by collecting resistivity data along two electrical profiles, each of which was 11.5 m long, with 24 steel electrodes with 0.5 m spacing. The two electrical profiles were set up, perpendicular to each other, with the same length and the same electrode spacing for each direction (x, y), and in a symmetrical position with respect to the center of the ring infiltrometer (Fig. 5). The so-called Wenner array was chosen as at the Altamura test site. The same instrument resistivity meter, IRIS–SYSCAL Switch Pro 48, was used to measure the resistivity of the subsurface. During the infiltrometer tests, the monitoring of the infiltrated water was carried out by means of a time-lapse technique, collecting more than 2,000 measurements for each acquisition. The total acquisition time was about 3 hours for the constant head test, and about 2.5 hours for the falling head test.

RESULTS

In the following paragraphs the results related to the two different test sites are described: the Altamura site consisting of limestone with and without visible fractures, such as an example of hard sedimentary rock, and the San Pancrazio site consisting of calcarenite, such as an example of soft sedimentary rock. Different approaches are used for the elaboration of the experimental data collected at the two geological outcrops. It is highlighted that in the case of fractured limestone the interpretation of the data collected experimentally, needs a numerical elaboration supported by mathematical models.

Limestone

The results of the infiltrometer tests are summarized in the graph in Figure 6, which shows a constant decrease in the water level in the ring during the tests, after the water level had reached its maximum depth of 0.13 m in both cases. The slope of the trend line of experimental sites gives an average infiltration rate equal to 1.33⍰0.0034 m d^{-1} for test #1, and 0.22⍰0.0027 m d^{-1} for test #2, where the standard deviations take into account the standard errors of the best fit procedures. In order to interpret the results of the falling head infiltrometer test, the simplified equation from Nimmo et al. (2009) was considered:

$$K_{fs} = \frac{L_G}{t}\ln\left(\frac{L_G+\lambda+D_o}{L_G+\lambda+D}\right) \qquad (5)$$

where Kfs (LT^{-1}) is the field-saturated hydraulic conductivity, t is the time, Do and D are the initial and final ponded depths, respectively, LG = C1d + C2b is the ring-installation scaling length, with 0.993 and 0.578 as recommended values for C1 and C2, while d and b are the ring insertion depth and ring radius, respectively, and ⍰⍰is an index of how strongly water is driven by capillary forces in a particular medium. The macroscopic capillary length, ⍰, during downward water movement in a fracture plane, is related to the fracture aperture and the interfacial surface tension ⍰⍰(M/T^2) between air and water (equal to 71.97 dyn/cm at 25°C) (de Gennes et al., 2002), using the following Young-Laplace equation (Pruess & Tsang, 1990):

$$\lambda = 2\frac{\sigma}{\rho g \times \bar{b}_c} \tag{6}$$

assuming that the water-air contact angle equals zero, and that the cutoff aperture equals the fracture aperture which delimits the region occupied only by the non-wetting phase, g (L/t^2) is the gravity acceleration and ⍰⍰ (M/L^3) is the water density. In order to estimate the field-saturated hydraulic conductivity value, Kfs, by means of equation (5), the macroscopic capillary length, ⍰, is required. It is determined by inverse modeling, using an unsaturated fractured flow model in the vertical 2-D fracture (Masciopinto & Benedini, 1999). The model solutions were then calibrated on the basis of the comparison between the shapes of the wetting front obtained from the simulation outputs and those of the electrical resistivity images at the correspondent time (Figs. 7 - 8). After estimating the macroscopic capillary length (0.90 m and 0.95 m for tests #1 and #2, respectively) and the corresponding cutoff aperture (16.5 ⍰m and 15.5 ⍰m), equation (5) gave field-saturated conductivity values of 0.67⍰0.01 m d^{-1} and 0.054⍰0.001 m d^{-1} for tests #1 and #2, respectively.

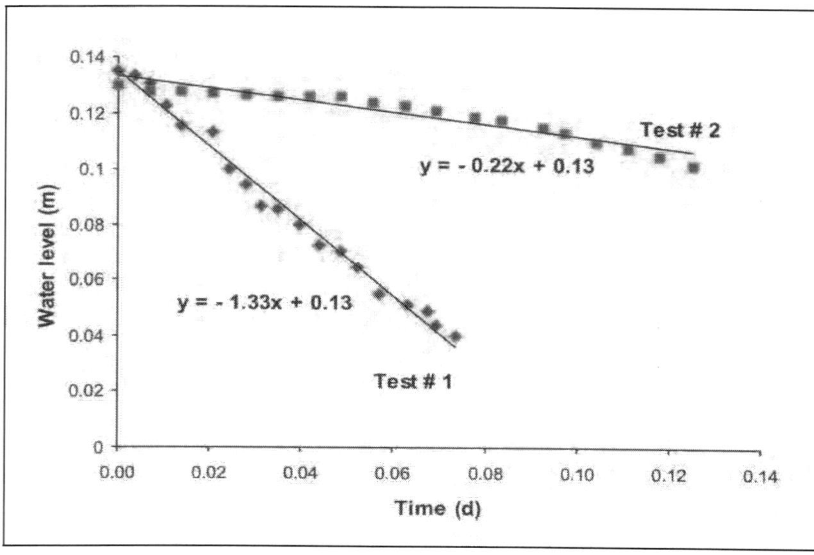

Fig. 6. Measured water level versus time during tests # 1 and #2 at the Altamura site.

The notable difference between the field-saturated hydraulic conductivity values of tests #1 and #2 highlights the great variability in hydraulic properties which characterizes the carbonate rocks studied. In fact, even though the two test sites were only 300 m apart, and the experiments were performed on the same geological formations, the infiltration rates observed in the areas tested were very different. For the Altamura fractured limestone, the modified Kozeny-Carman equation (Pape et al., 1999) leads to outcrop saturated conductivity values ranging from 0.6 m d^{-1} to 12 m d^{-1}, considering the rock outcrop porosity from 1% to 2%.

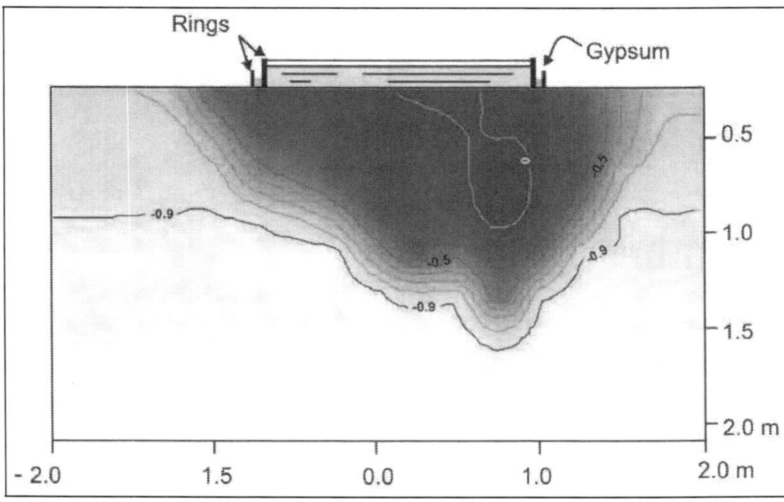

Fig. 7. Best selected model simulation output in order to fit the shape of the wetting front at t = 80 min. Contour lines represent model output expressed as matric potential (m) in the fracture.

It should be noted that the former conductivity value is very similar to the field-saturated conductivity obtained by the ring infiltrometer in test #1 (0.67 m d^{-1}). In contrast, the field-saturated conductivity (0.054 m d^{-1}) derived from test #2, where the limestone outcrop had no visible fractures, is close to the value (0.02 m d^{-1}) obtained by laboratory tests on the rock matrix of the Murge limestone (Borgia et al., 2002). The above comparison supports the field-saturated hydraulic conductivity values obtained using the large ring infiltrometers. Simultaneous electrical subsurface measurements, inverted by using the Res2Dinv software, also show a difference in behavior between the two tests as a consequence of the water infiltration during the tests (Fig. 8). Before the infiltrometer tests (t□0 min), the subsurface resistivity was above 500 □m at the first site and more than 2,000 □m at the second site. In test #1, the low resistivity area, below 150 □m, was visualized after 80 minutes to about 1.5 m of depth; similar results were obtained for the electrode arrays in the perpendicular direction. The low resistivity zones can be related qualitatively to fractured rock with high water content and moderate salinity. The same resistivity value was visualized at shallower depth, at the same time, during test #2 than the previous test. In fact, in the last case, it is not possible to evaluate the depth of wetting front from a quantitative point of view, because the resolution of the resistivity images depends on the electrode spacing. As rule of thumb, a structure or object in the subsurface, having dimensions less than the electrode spacing, cannot be defined clearly. For both tests the electrical resistivity measurements confirmed the effectiveness of the ring sealing, showing that the water infiltrated from within the ring and not from outside (Fig. 8). Additionally, electrical resistivity imaging highlights not only that the water reached different depths in the two infiltrometer tests, but also that its variable redistribution, within the investigated vertical plane, results from the structural characteristics of the limestone outcrop. As expected, in test #1, the deepening of the level of the low resistivity anomaly associated with the infiltration of moderately saline water was much greater than in test #2, and was, in fact, consistent with the different number and size of the fractures observed visually in the limestone.

Calcarenite

The results of the field experiments carried out on the calcarenitic layer of the vadose zone outcropping at the bottom of the quarry, are shown in Figures 9 and 10. The Figures show the infiltration data and the electrical resistivity images, respectively. In order to interpret the results of constant head infiltrometer test we have plotted the infiltration rate, q ?? dI / dt , versus time, t, as shown it the graph of Figure 9a. The value of q, initially decreases rapidly with time and then approaches a constant value. Practically, the rate of infiltration falls, starting from very high values at the beginning of the test when the subsurface is in an unsaturated condition, up to a value of $9 \times 10^{-4} \pm 0.0001$ cm/s (i.e. between 0.86 md^{-1} to 0.69 md^{-1}), when the rock is in quasi-saturated condition. For the studied calcarenite, this value of the infiltration rate, obtained by considering the horizontal asymptote of the curve, was reached after about 2 hours from the start of the infiltration, depending upon the water content of the rock at the beginning of the test. This value should be approximately equal to the field-saturated hydraulic conductivity value, Kfs. Again for the falling head infiltrometer test, equation (5) was considered. In this case a ⍰⍰ value equal to

0.01 m was used for the calcarenite, taking into account that the sensitivity of conductivity calculations to the value of ⍰⍰ is small (Nimmo et al., 2009), and that Elrick et al. (1989) proposed a ⍰⍰ value of about 0.08 m suitable for most soils with structural development, a value of 0.03 m for gravelly soils, and 0.25 m for fine-textured soil without macropores. Figure 9b shows the plot of the effective infiltration length (right-hand side of equation (5), except the variable t) vs. time and the slope of this regression line is a convenient calculation of Kfs. The graph shows how the Kfs values decrease during the infiltration test, from the highest value of 1.22x10^{-3} cm/s (i.e. 1.05 md^{-1}), at the start of the test, until the lowest of 5.61x10^{-4} cm/s (i.e. 0.48 md^{-1}) at the end of the infiltration, with an average value of 8.13x10^{-4} cm/s (i.e. 0.77 md^{-1}). This last value is consistent with the quasi-steady infiltration rate value, obtained from the constant head test (Fig. 9a). The variability of the field-saturated hydraulic conductivity, Kfs, observed in Figure 9b depends on the change in water content with the time during the experiment: higher Kfs values are obtained in the initially dry rock than in that wet. In fact, the soil water content during the test has been considered an important factor that affects the Kfs value of the experimental area (Bagarello & Sgroi, 2007). In order to have a rough estimation of the wetting front depth, a porosity value of about 0.4 was considered. Using this value of porosity and the high infiltration rate of about 4 cm/s for the first 20 minutes of the constant head test, it was estimated that the wetting front should reach a depth of about 20 cm, while up to the end of the falling head test, by considering

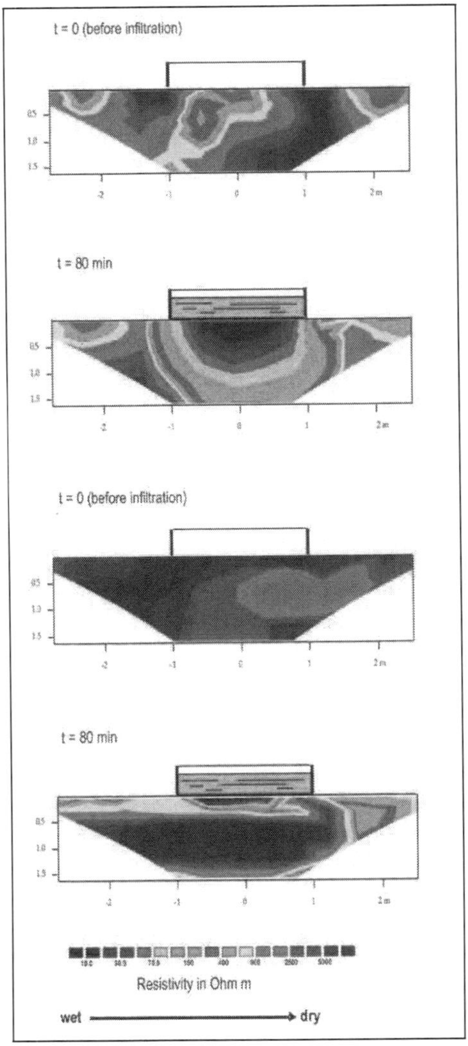

Fig. 8. Electrical resistity profiles during test # 1 and # 2, before (t = 0 min) and during (t = 80 min) water infiltration tests at the Altamura site.

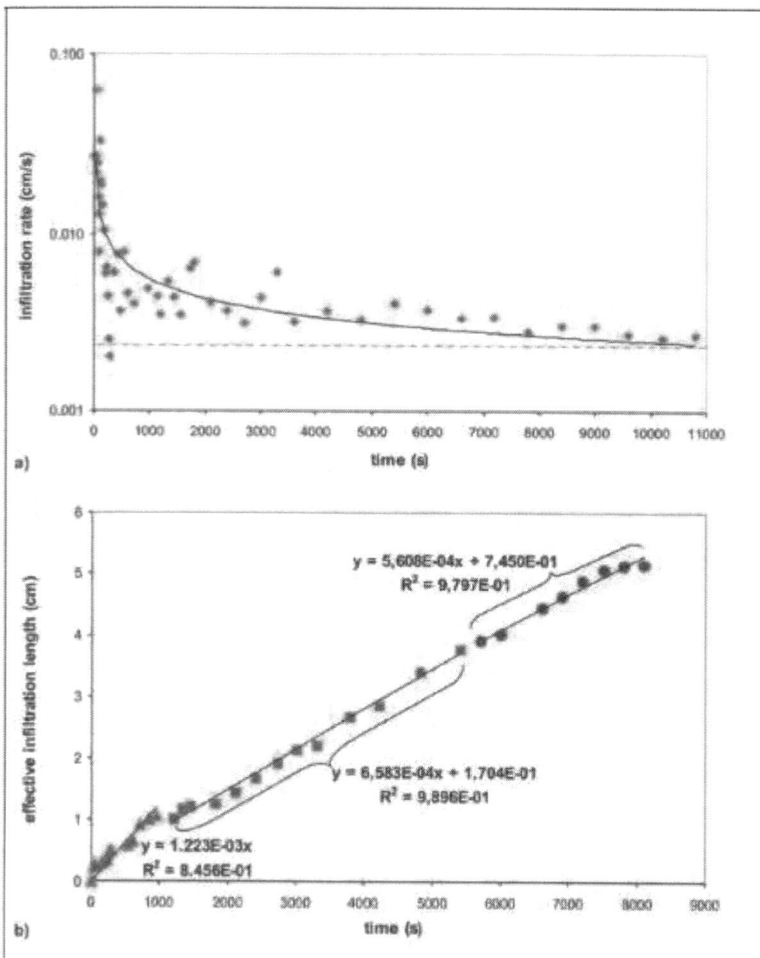

Fig. 9. Test at the San Pancrazio site: a) constant head infiltration test – infiltration rate versus time; b) falling head infiltration test – effective infiltration length versus time.

an average infiltration rate of about $9x10^{-4}$ cm/s, it should only deepen by about another 6 cm. Obviously, this so small a difference in depth that it is not perceptible in the chronological sequences of electrical resistivity images (Fig. 10).

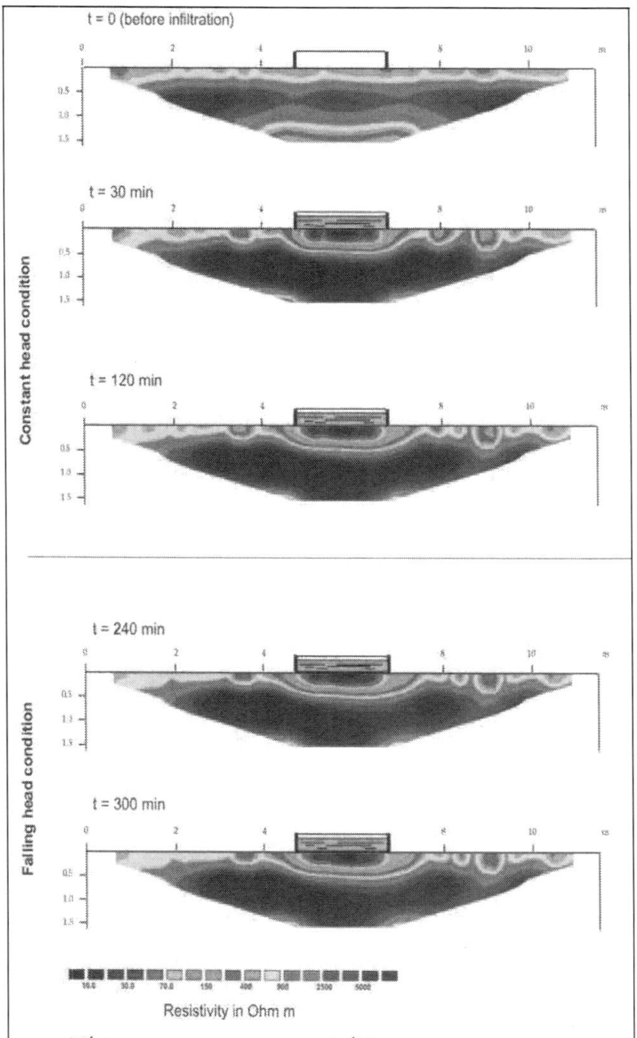

Fig. 10. Electrical resistivity profiles during infiltration test

at the San Pancrazio site – profile 1.

Knowing that the shallow aquifer is at 12 m below the bottom of the quarry where the waste was disposed, it was estimated that the time required for the pollutants to reach the shallow groundwater is just over 15 days. Concerning the results of the electrical resistivity surveys, using time-lapse techniques, 2-D resistivity images were obtained for different acquisition times. The electrical resistivity images showed no significant differences between the two perpendicular directions (Fig. 10). The first two profiles on the top of Figure 10, measured before starting water infiltration (t ⍰⍰0 min) and characterized by resistivity values above 1,000 ⍰m, represent the profiles with which to compare the following profiles, in order to monitor the water flux in the subsurface. The images from the constant head test show that the wetting front (resistivity values around 60 ⍰m) deepens in the first 20 minutes of the experiment, much more than in the following 4.5 hours, up to the end of the test. These results are consistent with the high values of infiltration rate, correspondent to the initial part of the constant head test. The images related to the sequences of the electrical resistivity profiles, acquired during the falling head test (Fig. 10), are similar to the last profile of the constant head condition, such as we expected by considering that the field-saturated hydraulic conductivity values, obtained from the tests under the different conditions, are similar.

CONCLUSION

The large ring infiltrometer, coupled with subsurface electrical resistivity measurements, described in this chapter, has proved to be a simple and inexpensive field tool, capable of evaluating the field-saturated hydraulic conductivity of rock formations, even if it requires numerical elaboration supported by mathematical models, in certain conditions. It is designed to be installed directly on an outcrop of both hard and soft rock, and easily constructed on site with inexpensive, lightweight materials, and has an adjustable diameter. These characteristics improve the versatility of the infiltrometer method and its adaptability to various geological conditions. Thus it expands the potential for exploring field-hydraulic conductivity of rocks, which until now has been investigated more frequently in laboratories, mainly because of the practical difficulties involved in field investigations. Simultaneous electrical resistivity measurements are used to monitor the subsurface water infiltration instead of humidity sensors or tensiometers, thus sidestepping the difficulty of inserting probes into the rock. Further technical difficulties, related to the installation in the field of the large ring infiltrometer, were solved using non-commercially available equipment during the on-site installation procedure. Specifically, the setup of the experimental apparatus requires the hollowing out of a furrow in which to install the ring and seal it to the rock surface with gypsum. Otherwise, if the field conditions of the infiltration area are very heterogeneous, consisting of different media, e.g. rock and soil, an infiltrometer made of two concentric rings is suggested in order to improve the hydraulic packing during the water infiltration test. In this case the continuity between the ring infiltrometer wall and the rock-soil surface is obtained by filling the space between the external and internal rings with gypsum, up to a height of about 2.0 cm,to create a seal. Thus the second ring improves the gypsum seal of the first ring with the ground, overcoming the challenge presented by the presence of different media along the edge of the ring. The efficacy of the seal of ring infiltrometer with soil/rock surfaces was confirmed by the simultaneous electrical resistivity measurements that show the deepening and spreading of the water during the infiltration tests. The time required for the installation was about two hours and the water volume used for each infiltration test was about 0.5 m^3 for the limestone, and about 0.3 m^3 for calcarenite, depending on the local rock permeability. On the whole, the field- hydraulic conductivity data obtained from the infiltrometer tests (0.67 m d^{-1} and 0.054 m d^{-1}, for tests #1 and #2 on limestone, respectively, and 0.77 m d^{-1} for calcarenite) are consistent with the nature of the rocks tested and are corroborated by laboratory measurements carried out by other authors (Quarto & Schiavone, 1994; Borgia et al., 2002). The difference between the field-saturated hydraulic conductivity values obtained by the infiltration tests, highlights the difference of the outcrops studied, owing to the different geological formations characterized from variable heterogeneity degree (number and size of fractures). The possibility of building a ring with a large adjustable diameter on site, has the advantage of including fractures and other features of the rock in order to obtain field-saturated hydraulic conductivity data at a more representative scale of measurements, with negligible border effects. The proposed installation procedure of a large ring infiltrometer on rocky outcrops extends the possibility of performing field infiltrometer tests on rocks that would, previously, have been very difficult to test. Simultaneous subsurface electrical resistivity imaging, using time-lapse techniques, is a useful indicator of water infiltration/redistribution. Moreover, the electrical resistivity survey provides evidence that the conductivity structure is confined under the ring infiltrometer, confirming that the apparatus works well in order to minimize the lateral lack of water, without having losses along the edge of the ring, allowing the acquisition of accurate and representative experimental data. On the other hand, it is important to highlight that the geophysical inversion model used, Res2Dinv, has not been able to represent the heterogeneities and anisotropies of the heterogeneous system, such as fractured or karstic rocks. This consideration encourages further research and study in order to define new algorithms able to derive real images closer to the real subsurface features than now, starting from the electrical resistivity measurements of heterogeneous systems. Using the simplified equation for determining the field-saturated hydraulic conductivity from the experimental data of falling head infiltrometer tests, it is demonstrated that the proposed method works well, in different cases, proving it is a good tool to know how quickly the water moves through the unsaturated zone in real field conditions.

ACKNOWLEDGMENT

The work was partially funded by the Italian Government (Regional Autority), which is gratefully acknowledged. We thank Costantino Masciopinto for his collaboration in modelling on fractured systems and for his fruitful cooperation during the field experiments. We also thank Rita Masciale and Francesco De Benedictis for their support for the field work and for their comments during the preparation and revision of this document.

REFERENCES

1. Archie, G.E. (1942). The Electrical Resistivity Log as an Aid in Determining Some Reservoir Characteristics. *Petroleum Transactions of AIME*, Vol. 146, pp. 54–62.

2. Bagarello, V. & Sgroi, A. (2007). Using the Simplified Falling Head Technique to Detect Temporal Changes in Field-Saturated Hydraulic Conductivity at the Surface of a Sandy Loam Soil. *Soil & Tillage Research,* Vol. 94, pp. 283–294.

3. Baumhardt, R.L., Lascano, R.J. & Evett, S.R. (2000). Soil Material, Temperature, and Salinity Effects on Calibration

of Multisensor Capacitance Probes. *Soil Science Society of America Journal*, Vol.64, pp. 1940-1946.

4. Binley, A.M., Cassiani, G., Middleton, R. & Winship, P. (2002). Vadose Zone Flow Model Parameterization Using Cross-Borehole Radar and Resistivity Imaging. *Journal of Hydrology*, Vol. 267 (3-4), pp. 147-159.

5. Bogena, H.R., Huisman, J.A., Oberdosrster, C.A. & Vereecken, H. (2007). Evaluation of a Low-Cost Soil Water Content Sensor for Wireless Network Application. *Journal of Hydrology*, Vol. 344, pp. 32-42.

6. Borgia, G.C., Bortolotti. V. & Masciopinto, C. (2002). Valutazione del Contributo della Porosità Effettiva alla Trasmissività di Acquiferi Fratturati con Tecniche di Laboratorio e di Campo (Evaluation of Effective Porosity Contribution to the Transmissivity of Fractured Aquifer Using Laboratory and Field Techniques). *IGEA, Groundwater Geoengineering*, Vol. 17, pp. 31-43.

7. Cassiani, G., Bruno, V., Villa, A., Fusi, N. & Binley, A.M. (2006). A Saline Tracer Test Monitored via Time-Lapse Surface Electrical Resistivity Tomography. *Journal of Applied Geophysics*, Vol. 59, pp. 244-259.

8. Castiglione, P., Shouse, P.J., Mohanty, B., Hudson, D. & van

Genucthen, M.Th. (2005). Improved Tension Infiltrometer for Measuring Low Flow Rates in Unsaturated Fractured Rock. *Vadose Zone Journal*, Vol. 4, pp. 885-890.

9. Ciaranfi, N., Pieri, P. & Ricchetti, G. (1988). Note alla Carta Geologica delle Murge e del Salento (Puglia Centro-Meridionale) (Notes on the Geological Map of Murge and Salento (Central-Southern Puglia). *Memorie.della Società Geologica Italiana,* Vol. 41, pp. 449-460.

10. Dahlin, T. (2001). The Development of DC Resistivity Imaging Techniques. *Computers & Geosciences,* Vol. 27, pp. 1019-1029.

11. Dahlin, T. (1996). 2-D Resistivity Surveying for Environmental and Engineering Applications. *First Break*, Vol. 14 (7), pp. 275–283.

12. Daily, W., Ramirez, A., LaBrecque, D. & Nitao, J. (1992). Electrical Resistivity Tomography of Vadose Water Movement. *Water Resources Research*, Vol. 28(5), pp. 1429-1442.

13. de Gennes, P.G., Brochard-Wyart, F. & Quéré, D. (2002). Capillary and Wetting Phenomena: Drops, Bubbles, Pearls, Waves. Springer. ISBN 0-387-00592-7.

14. deGroot-Hedlin, C. & Constable, S. (1990). Occam's

Inversion to Generate Smooth, Two- Dimensional Models from Magnetotelluric Data. *Geophysics*, Vol. 55, pp. 1613-1624.

15. Deiana, R., Cassiani, G., Kemna, A., Villa, A., Bruno, V. & Bagliani, A. (2007). An Experiment of Non-Invasive Characterization of the Vadose Zone via Water Injection and Cross-Hole Time-Lapse Geophysical Monitoring. *Near Surface Geophysics*, Vol. 5, pp. 183-194.

16. Edwards, L.S. (1977). A Modified Pseudosection for Resistivity and Induced-Polarization. *Geophysics,* Vol. 42, pp. 1020-1036.

17. Elrick, D.E, Reynolds, W.D. & Tan, K.A. (1989). Hydraulic Conductivity Measurements in the Unsaturated Zone Using Improved Well Analyses. *Ground Water Monitoring Review*, Vol. 9, pp. 184-193.

18. Griffiths, D.H. & Barker, R.D. (1993). Two-Dimensional Resistivity Imaging and Modelling in Areas of Complex Geology. *Journal of Applied Geophysics*, Vol.29, pp. 211-226.

19. Jones, S.B. & Or, D. (2004). Frequency Domain Analysis for Extending Time Domain Reflectometry Water Content Measurement in Highly Saline Soils. *Soil Science Society of America Journal*, Vol. 68, pp. 1568–1577, 2004.

20. Kizito, F., Campbell, C.G., Cobos, D.R., Teare, B.L., Carter, B. & Hopmans, J. W. (2008). Frequency, Electrical Conductivity and Temperature Analysis of a Low-Cost Capacitance Soil Moisture Sensor. *Journal of Hydrology*, Vol. 352, pp. 367-378.

21. Lai, J. & Ren, L. (2007). Assessing the Size Dependency of Measured Hydraulic Conductivity Using Double-Ring Infiltrometers and Numerical Simulation. *Soil Science Society of American Journal,* Vol. 71, pp. 1667-1675.

22. Leeds-Harrison, P.B., Youngs, E.G. & Uddin, B. (1994). A Device for Determining the Sorptivity of Soil Aggregates. *European Journal of Soil Science,* Vol. 45, pp. 269-272.

23. Loke, M.H. (2001). *A tutorial on 2-D and 3-D electrical imaging surveys.* Avalaible from www.geoelectrical.com.

24. Loke, M.H. & Barker, R.D. (1996). Rapid Least-Squares Inversion of Apparent Resistivity Pseudosections by a Quasi-Newton Method. *Geophysical Prospecting,* Vol. 44, pp. 131-152.

25. Masbrucha, K. & Ferré, T.P.A. (2003) A Time-Domaine Transmission Method for Determining the Dependence of the Dialectric Permittivity on Volumetric Water Content. An Application to Municipal Landifills. *Vadose Zone Journal,* Vol. 2, pp. 186-192.

26. Masciopinto, C. (2005). Pumping-Well Data for Conditioning the Realization of the Fracture Aperture Field in Groundwater Flow Models. *Journal of Hydrology*, Vol. 309 (1-4), pp. 210-228.

27. Masciopinto, C. & Benedini, M. (1999). Unsaturated Flow in Fractures with Anisotropic Variable Apertures, *Proceedings of XXVIII IAHR - AIRH Congress*, Graz, Austria, August 23–27, 1999.

28. Nimmo, J.R., Schmidt, K.S., Perkins, K.S. & Stock J.D. (2009) Rapid Measurement of Field- Saturated Hydraulic Conductivity for Areal Characterization. *Vadose Zone Journal*, Vol 8, pp.142-149.

29. Pape, H., Clauser, C. & Iffland, J. (1999). Permeability Prediction Based on Fractal Pore- Space Geometry. *Geophysics*, Vol. 64(5), pp. 1447-1460.

30. Pruess, K. & Tsang, Y.W. (1990). On Two-Phase Relative Permeability and Capillary Pressure on Rough-Walled Rock Fractures. *Water Resources Research*, Vol. 26(9), pp. 1915-1926.

31. Quarto, R. & Schiavone, D. (1994). Hydrogeological Implications of the Resistivity Distribution Inferred from Electrical Prospecting Data from the Apulian Carbonate Platform. *Journal of Hydrology*, Vol. 154, pp. 219-244.

32. Reynolds, J. M. (1998). *An Introduction to Applied and Environmental Geophysics*, Wiley, (Ed.) John Wiley and Sons, ISBN 0-471-9555-8, New York.

33. Reynolds, W.D., Elrick, D.E. & Young, E.G. (2002). Ring or Cylinder Infiltrometers (Vadose Zone), In: *Method of Soil Analysis,* J.K. Dane & G.C. Topp, (Eds.), pp. 818-826, Soil Science Society of America, Inc., ISBN 0-89118-841-X Madison, Wisconsin, USA.

34. Robinson, D.A., Jones, S.B., Wraith, J. M., Or, D. & Friedman, S. P. (2003). A Review of Advances in Dielectric and Electrical Conductivity Measurement in Soils Using Time Domain Reflectometry. *Vadose Zone Journal,* Vol. 2, pp. 444– 475.

35. Seyfried, M.S., & Murdock, M.D. (2004). Measurement of Soil Water Content with a 50-Mhz Soil Dielectric Sensor. *Soil Science Society of America Journal,* Vol. 68(2), pp. 394-403.

36. Swartzendruber, D., & Olson, T.C. (1961). Model Study of the Duble Ring Infiltrometer as Affected by Depth of Wetting and Particle Size. *Soil Science,* Vol. 92, pp. 219-225.

37. Telford, W.M., Geldart, L.P. & Sheriff, R.E. (1990). *Applied Geophysics* (2nd edition), Cambridge University Press, ISBN 0-521-33938-3, Cambridge.

38. Youngs, E.G., Spoor, G. & Goodall, G.R. (1996). Infiltration from Surface Ponds into Soils Overlying a Very Permeable Substratum. *Journal of Hydrology*, Vol. 186, pp. 327–334.

39. White, P.A. (1988). Measurement of Ground-Water Parameters Using Salt-Water Injection and Surface Resistivity. *Ground Water*, Vol. 26, pp. 179–186.

Index

N

O

R

T

V

W